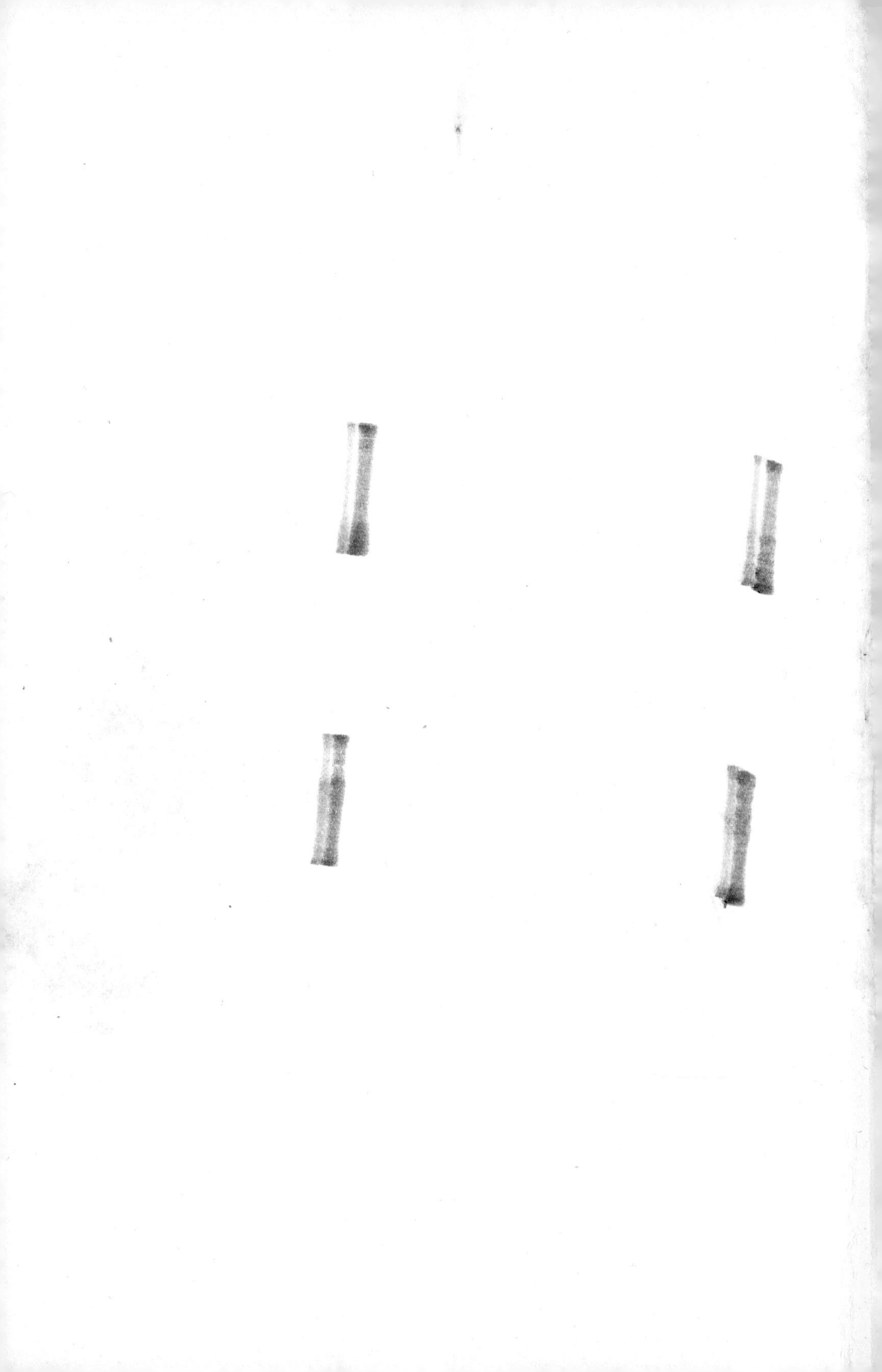

SUPERCONDUCTING ELECTRON-OPTIC DEVICES

THE INTERNATIONAL CRYOGENICS MONOGRAPH SERIES

H. J. Goldsmid
Thermoelectric Refrigeration, 1964

G. T. Meaden
Electrical Resistance of Metals, 1965

E. S. R. Gopal
Specific Heats at Low Temperatures, 1966

M. G. Zabetakis
Safety with Cryogenic Fluids, 1967

D. H. Parkinson and B. E. Mulhall
The Generation of High Magnetic Fields, 1967

W. E. Keller
Helium-3 and Helium-4, 1969

A. J. Croft
Cryogenic Laboratory Equipment, 1970

A. U. Smith
Current Trends in Cryobiology, 1970

C.A. Bailey
Advanced Cryogenics, 1971

D. A. Wigley
Mechanical Properties of Materials at Low Temperatures, 1971

C. M. Hurd
The Hall Effect in Metals and Alloys, 1972

E. M. Savitskii, V. V. Baron, Yu. V. Efimov, M. I. Bychkova, and L. F. Myzenkova
Superconducting Materials, 1973

W. Frost
Heat Transfer at Low Temperatures, 1975

I. Dietrich
Superconducting Electron-Optic Devices, 1976

SUPERCONDUCTING ELECTRON-OPTIC DEVICES

I. Dietrich

Research Laboratories
Siemens A.G.
Munich, West Germany

PLENUM PRESS • NEW YORK AND LONDON

Library of Congress Cataloging in Publication Data

Dietrich, I

Superconducting electron-optic devices.

(The International cryogenics monograph series)
Bibliography: p.
Includes index.
1. Electron microscope. 2. Superconductivity. I. Title. II. Series
QH212.E4D53 502'.8 76-20466
ISBN 0-306-30882-7

A Division of Plenum Publishing Corporation
227 West 17th Street, New York, N.Y. 10011

Printed in the United States of America

* | Preface

Electron optics involves the influence of electric and magnetic fields on electron beams. In those electron optical instruments utilizing magnetic fields, a replacement of the conventional, i.e., nonsuperconducting, electron optical parts, is worth considering if the outstanding magnetic properties of superconductors can improve the systems. However, the use of superconductors demands complicated cryogenic techniques and this, of course, dampens enthusiasm. There are fields, however, where there are extreme requirements on the optical systems, namely, electron microscopy and high-energy physics. The great advantage of the combination of electron optics and superconductivity in these domains has been demonstrated in recent experiments.

This monograph is mainly concerned with electron microscopy. Superconductivity in high-energy electron optics is treated only briefly, in Appendix A, since the author is little acquainted with the details of the projects. Furthermore, the number of experiments as yet carried out is small. In Appendix B, electron microscope studies of basic superconductor phenomena are reviewed. This material is included, even though it is only slightly connected with the main topic of the book, since a breakthrough in this field may be possible by the application of superconducting lenses.

It is a pleasure for me to thank the management of the Research Laboratories of the Siemens Company for giving me the opportunity to write this book. I am most grateful to my colleagues at Siemens: F. Fox, G. Lefranc, E. Knapek, K. Nachtrieb, R. Weyl, and H. Zerbst, who are working on the project "Superconducting Lenses" and who supported me in every respect. I am very much indebted to K.-H. Herrmann, Fritz Haber Institut Berlin, for many discussions and advice, and to C. Passow, Kernforschungszentrum Karlsruhe, for allowing me to use unpublished material. I would also like to thank M. Kaczmarek for typing the manuscript, H. Yükselgil for preparing the drawings, and K.-L. Rau and J. M. Drew for their help in revision of the text.

I. Dietrich

Contents

1 Historical Survey 1

2 Basic Principles of Electron Optics 5

2.1. Rotationally Symmetric Lenses in the Bell-Shaped Field Approximation 5
2.2. Rotationally Symmetric Lenses with Arbitrary Field Distribution 13
2.3. Aberrations Resulting from Misalignment 14
2.4. Multipole Fields for Beam Correction 15
2.5. Image Contrast 17
2.6. Further Sources of Error 19
2.7. Fixed Beam and Scanning Mode 19

3 Superconducting Devices in Electron Microscopy 23

3.1. Advantages of Superconducting Devices 23
3.1.1. Improvements Resulting from the Use of Superconducting Materials 23
3.1.2. Improvements Resulting from the Liquid Helium Temperature in the Optical Devices 25
3.1.3. Economic Aspects 26
3.2. Technical Problems 26
3.2.1. Construction of Cryostats 26

3.2.2. Superconducting Materials 31
3.2.3. Normal Materials 40

4 Lens Design and Testing 45

4.1. Lens Design and Field Distribution 45
4.1.1. General Remarks 45
4.1.2. Coil Lens 46
4.1.3. Ring Lens 49
4.1.4. Iron Circuit Lens...................... 53
4.1.5. Shielding Lens 61
4.1.6. Advantages and Drawbacks of the Different Lens Types 64
4.2. Correction Systems for Superconducting Objective Lenses 66
4.3. Testing of Objective Lenses 69
4.3.1. Testing Objective Lenses with Simulation Devices 69
4.3.2. Testing in the Microscope 74

5 Systems with Superconducting Lenses 77

5.1. Tested Systems 77
5.2. Projected Systems 89

6 Other Superconducting Elements for Electron Microscopy 91

6.1. Superconducting High-Voltage Beam Generator 91
6.2. Magnetic Dipoles 95

7 Proposed Superconducting 3-MV Microscope .. 97

7.1. Accelerator 97
7.2. Spectrometer 104
7.3. Microscope Column 106
7.4. Further Improvements of the System 112

Appendixes 115

A. Superconducting Electron Optical Systems for High-Energy Physics 115

A.1. General Remarks 115
A.2. Magnet Designs 117

B. Application of Electron Microscopy to Basic Research on Superconductivity 120

B.1. Imaging by the Decoration Method 122
B.2. Imaging by Electron Shadow Microscopy 124
B.3. Imaging by an Electron Mirror Microscope 125
B.4. Imaging by Lorentz Microscopy 126
B.5. Imaging by a Vortex Electron Microscope 127

References 129

Index 137

1 | Historical Survey

The possibility of applying the phenomena of superconductivity to electron optical lenses was probably first proposed by Boersch in 1934 at the Berlin Laue colloquium, after the presentation of Meissner's paper on field repulsion in superconductors. Because the superconductors known at that time exhibited relatively low critical temperatures and small critical current densities at high fields, the problem was not further pursued. After the discovery of the high-temperature superconductors NbN and NbC [a transition temperature above 20 K was published for these compounds as the result of a measurement error (Aschermann *et al.*, 1941)], Diepold and Dosse (1941) invented a superconducting lens. The patent specification describes a device in which the imaging effect is produced by a superconducting ring enclosed in a helium container. The excitation of the ring is produced by a normal-conducting iron circuit magnet inside the microscope. Also described is a magnetization device outside the microscope for exciting the ring. The cooled part of the apparatus, the magnetized superconducting ring in its helium container, has to be introduced into the microscope. A shim coil is added for focusing. Judging this invention with the knowledge acquired in the last 35 years, we see immediately that it could never work. However, it is interesting to note that many of the remarkable features of superconducting lenses had already been described at that time.

World War II stopped further development in this direction.

When, at the end of the 1950s, the high-field superconductors with relatively large critical current densities were discovered, the application of superconductors for electron microscope lenses and for devices in high-energy physics again became of interest. In 1960, a U.S. patent was filed (Buchold, 1961) which proposed the use of superconductors in rotationally symmetric electron microscope lenses with iron circuits. In this proposal, the normal-conducting coils are replaced by superconducting ones, and superconducting shielding devices are designed to force the magnetic field in such a direction that the axial field is increased and the stray field is diminished. Unfortunately, such a lens was never tested.

Experiments with superconducting lenses were first carried out in the mid-1960s (Boersch *et al.*, 1964; Laberrigue and Levinson, 1964). The first published micrograph achieved with a superconducting device was taken by Fernández-Móran (1965). He performed the imaging with a Nb–Zr solenoid (diameter 50 mm, length approximately 60 mm) without an iron circuit, operated at 3.2 T in the persistent-current mode. The beam voltage was only 4–8 kV in this experiment. The exceptional stability of the picture was noteworthy.

At about the same time, a group at the Technical University of Berlin (Boersch *et al.*, 1966*a*) obtained a focusable image with a current-carrying superconducting ring made of niobium and excited in the pole pieces of an iron-circuit lens. Other groups which had already established results were experimenting at Hitachi (Osaza *et al.*, 1966), the University of Paris (Levinson *et al.*, 1965; Laberrigue and Severin, 1967), at Cornell University (Siegel *et al.*, 1966; Kitamura *et al.*, 1966*a*), and at the Siemens Research Laboratories (Dietrich *et al.*, 1967; Dietrich and Weyl, 1968). Hardy (1973) gives a review of the work conducted on superconducting electron microscope lenses up to 1970.

In the early 1960s, the Stanford group (Fairbank *et al.*, 1963) suggested building a superconducting linear accelerator, and in 1964 electrons were accelerated with a superconducting radio-frequency structure for the first time (Schwettmann *et al.*, 1966). Electrons at 80 keV were injected into a lead-plated three-cell copper cavity. An energy of approximately 500 keV was obtained.

After this success, more pretentious projects were begun, and in 1968 the Stanford group began the design of a 36-MeV superconducting accelerator.

For normal and superconducting accelerators and other high-energy facilities, magnets are essential for beam focusing, beam deflection, beam transport, and energy selection. After the first dc 5-T superconducting solenoids had been built in 1961 and a 6.8-T solenoid in 1962 (Riemersma *et al.*, 1963), plans were made to replace the conventional large magnets by the superconducting version in high-energy physics and thus reduce the operating costs. Unfortunately, the superconducting material available at that time exhibited strong degradation, i.e., the critical data obtained in small samples could not be realized in large devices. This was an enormous handicap. After the discovery of the cryostatic stability criterion by Steckly and Zar (1965), composite superconductors based on fine Nb–Ti superconducting filaments embedded in a Cu matrix came on the market. With these materials, it became feasible to construct large beam-transport magnets with a maximum induction $B_{max} = 6$ T combined with an overall current density $i_{eff} = 2 \times 10^4$ A/cm^2. A comprehensive monograph on superconducting magnets was recently published by Brechna (1973).

2 | Basic Principles of Electron Optics

The electron optical relations summarized in this chapter are essentially oriented toward electron microscopy. The course of the beam in a conventional transmission electron microscope is sketched in Figure 2.1. The electrons emitted from the cathode are first accelerated in a system of electrostatic lenses. The beam then enters the condenser system, which adjusts the illuminated area of the specimen. The specimen is situated somewhat above the center of the objective lens. The magnified image of the specimen is produced by the objective lens. Further magnification is achieved by additional lenses. Contrast, i.e., variation of the intensity of the electron beam image, is accomplished by scattering of the electrons by the atoms of the specimen and/or as a result of the phase shift of the electron waves in the specimen.

2.1. Rotationally Symmetric Lenses in the Bell-Shaped Field Approximation

The most important property of electron microscope systems is the resolving power of the objective lenses. The resolving power depends on the image aberrations. For a discussion of the basic features of a magnetic lens, we assume an axially symmetric field. The lens effect is caused by the deflection of the electrons by the Lorentz force

$$\mathbf{v} \times \mu_0 \mathbf{H}$$

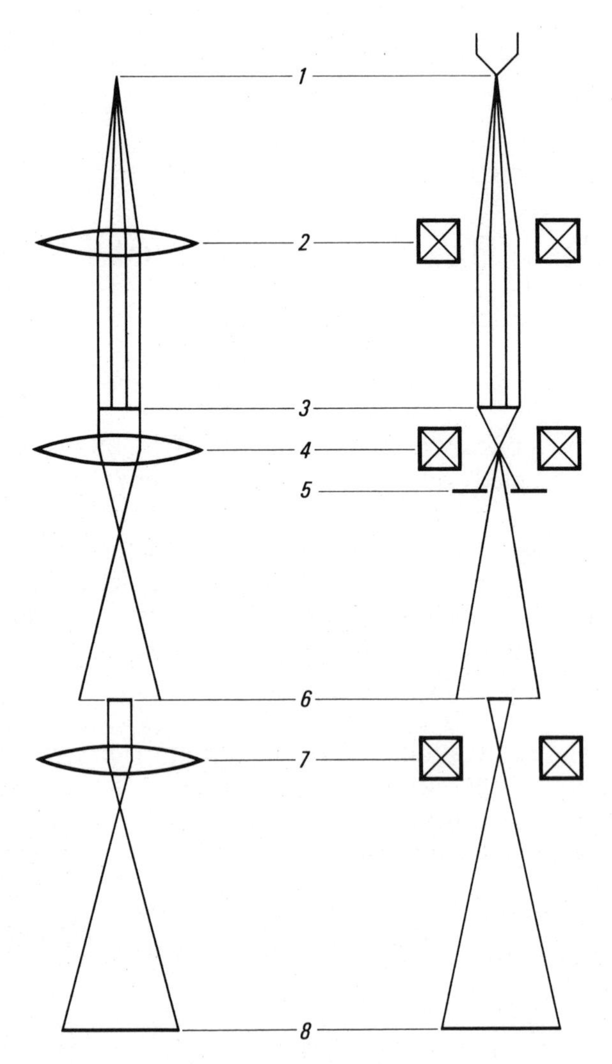

Figure 2.1. Light optical microscope (left) and fixed-beam transmission electron microscope: (1) electron source; (2) condenser lens; (3) specimen; (4) objective lens; (5) contrast aperture; (6) intermediate image; (7) projective lens; (8) image.

where **v** is the velocity of an electron and **H** is the field strength.

As the beam diameter is normally small and presumed to be perfectly aligned with the optical axis (z axis), interest centers on the axial field distribution H_z. As a first approach, H_z is assumed to be bell-shaped (Figure 2.2), corresponding to the relation (Glaser, 1952)

$$\frac{H_z}{H_0} = \frac{1}{1 + (z/d)^2} \tag{2.1}$$

where H_0 is the peak field and $2d$ is the half-width.

A lens is characterized by its strength k^2, which is a function of d, H_0, and the relativistically corrected beam voltage V^*:

$$k^2 = \frac{e}{8m_0} \frac{(\mu_0 H_0 \cdot d)^2}{V^*}$$

$$V^* = V\left(1 + \frac{e}{2m_0} \frac{V}{c^2}\right) \tag{2.2}$$

Here V is the beam voltage, e the electron charge, and m_0 the electron rest mass. Equation (2.2) can also be written as

$$k = 148 d\mu_0 H_0 / V^{*1/2} \qquad \text{in SI units} \tag{2.2'}$$

The curves in Figure 2.3 are plotted from equations (2.2) or (2.2′); they show d as a function of $\mu_0 H_0$ for various values of V at $k^2 = 1$ and $k^2 = 3$.

Lenses are called weak for $k^2 < 1$ and strong for $k^2 \geq 1.5$. For sufficiently strong lenses the location of some of the cardinal

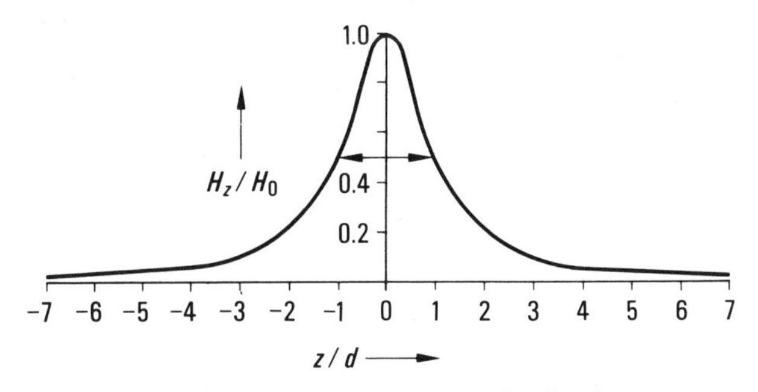

Figure 2.2. Bell-shaped field distribution.

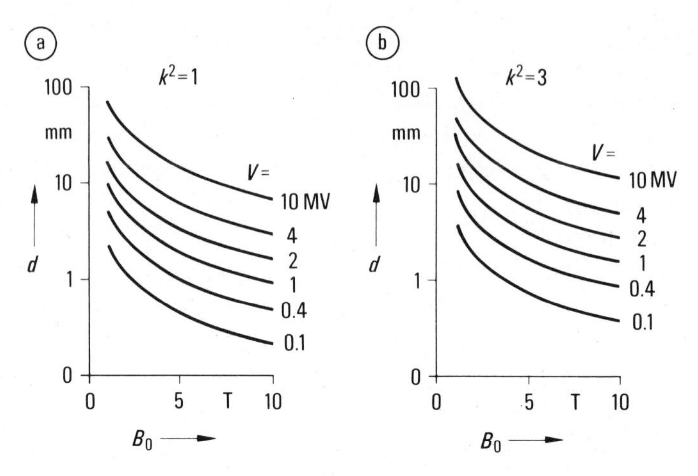

Figure 2.3. Graphical presentation of equation (2.2) for (a) $k^2 = 1$ and (b) $k^2 = 3$.

elements is determined by tracing rays (Liebmann and Grad, 1951) as in Figure 2.4, where two imaginary principal rays are used. The ray entering the lens parallel to the z axis is deflected toward the axis by the magnetic field. The intersection of this

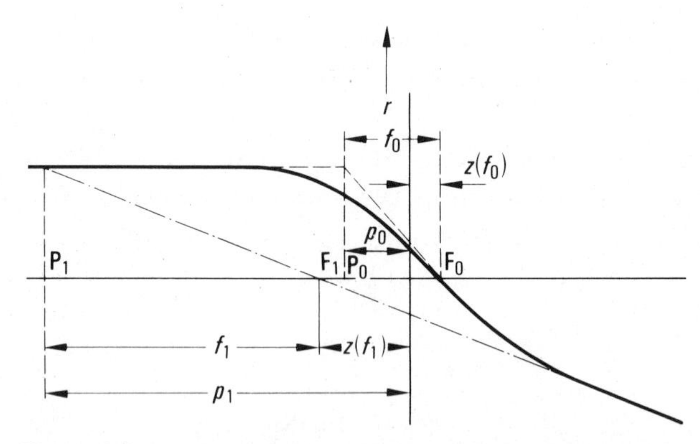

Figure 2.4. Ray-tracing diagram. P_0 and P_1 are the principal planes, and p_0 and p_1 are the corresponding distances of these from the midplane; F_0 and F_1 are the focal points, and f_0, f_1 are the corresponding focal lengths; $z(f_0)$ and $z(f_1)$ are the z coordinates of the focal points F_0 and F_1, respectively.

ray with the z axis is the focal point F_0, located a distance $z(f_0)$ from the midplane. The inclination of the electron trajectory after it passes F_0 is reduced by the ultrafocal field. The distance p_0 between the optical principal plane P_0 and the midplane is determined by the point of intersection of the tangent to the trajectory at F_0 and the extension of the initial line of the trajectory. The focal length for imaging a specimen situated near F_0 is given by $f_0 = z(f_0) + p_0$. Another focal point F_1 is obtained by the intersection of the z axis and the extension of the straight-line portion of the trajectory on the right side. The corresponding principal plane P_1 is determined by the intersection of the latter straight line and the initial line of the trajectory. The distance f_1 between P_1 and F_1 is called the asymptotic focal length; f_1 is useful when using the lens as a projector, i.e., if the electrons arrive from a very small source at a distance $\gg f_0$.

For the bell-shaped field we have

$$f_0 = \frac{d}{\sin[\pi/(1 + k)^{1/2}]}$$

$$f_1 = \frac{d(1 + k^2)^{1/2}}{\sin(1 + k^2)^{1/2}} \tag{2.3}$$

If the lenses are not too strong ($k^2 \approx 1$), the asymptotic focal length f_1 is approximately equal to f_0. For $k^2 = 2$, we have $f_1/f_0 = 2.5$, and for $k^2 = 3$ we have $f_0 = d$, $f_1 \to \infty$.

In electron microscopy strong lenses are usually used for the objective lenses since large magnification combined with a small length of the imaging system is desired. Furthermore, two important image errors, chromatic and spherical aberration, approach their relative minimum values for $k^2 = 3$ and $k^2 = 7$, respectively. With $k^2 = 1$, the specimen is situated in a region of small field strengths. If $1.5 < k^2 < 3$, the specimen is immersed so far in the forefield that the illumination parameter is already changed considerably by the forefield. At $k^2 \approx 3$, the beam crosses the optical axis in the center of the bell-shaped field, and the specimen is located in the midplane of the lens. In the so-called one-field-condenser mode at $k^2 \approx 3$, the forefield reduces the beam cross section so that the illuminated spot on the specimen

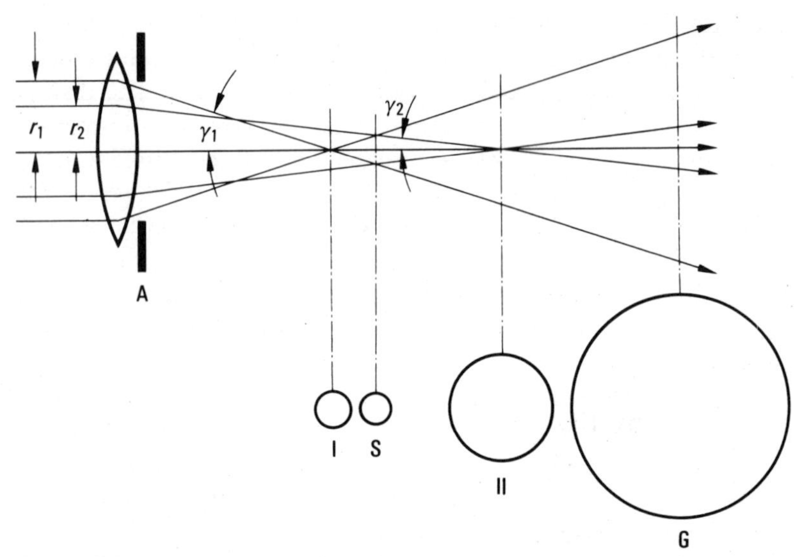

Figure 2.5. Principle of spherical aberration: (A) aperture; (I) focus of rays with radius r_1; (II) focus of rays with radius r_2; (G) focus of paraxial rays (Gauss focus); (S) plane of least confusion (Scherzer, 1949).

is very small. Besides other modes, this can be used in scanning transmission microscopy, which will be discussed later. The second part of the field has the effect of a magnifying lens. There are other modes with higher k^2 [for example $k^2 = 5$ (Suzuki, 1965)], which have been tested but are rarely applied.

The relationships for an approximate calculation of some of the image aberrations are very simple in the region $1.5 < k^2 < 5$ if we assume the specimen consists of small spots which emit electrons incoherently.

Spherical aberration occurs if we give up the paraxial approximation. The rays emitted from a point object are then imaged as a disk, the size of the disk varying with the distance along the optical axis, as shown in Figure 2.5. If the position of the smallest disk, the *disk of least confusion*, is considered as the new focal point, the focal length will vary with the size of the aperture. The change in focal length is proportional to the square of the aperture angle γ,

$$\Delta f_{0s} = c_s \gamma^2$$

where c_s is the spherical aberration constant and is given by

$$c_s = df_s(k^2) \tag{2.4}$$

where the function $f_s(k^2)$ varies little with k^2 and is of the order of magnitude of 1.

Chromatic aberration is caused by the following:

(1) the ripple and drift of the objective lens current;

(2) variations in the beam voltage, and the energy spread of the electrons due to the temperature of the cathode;

(3) the width of the conduction band from which the electrons are emitted;

(4) the interaction of electrons in the beam (Boersch effect).

The inelastic scattering of the electrons in the specimen produces a further spread in their energy distribution. These effects cause a broadening of the focal length. The broadening due to the energy spread of the electrons is given by

$$\Delta f_{0c} = c_c \frac{\Delta(eV^*)}{eV^*} \gamma \tag{2.5}$$

where eV^* is the electron energy, and c_c is the chromatic aberration coefficient:

$$c_c = df_c(k^2) \approx \tfrac{1}{2} df_s(k^2) \tag{2.6}$$

For $k^2 = 3$, we find

$$c_s = 0.3d \approx c_c/2$$

The focal length broadening due to variations in the objective lens current I is given by

$$\Delta f_{0c} = c_c \frac{2\Delta I}{I} \gamma \tag{2.7}$$

The resolution of the system is defined as the distance between two closely adjacent spots which are just resolved in the image plane. To obtain the theoretical limit of resolution δ_s, an optimum objective aperture angle γ_0 must be chosen so that the sum of the spherical aberration and diffraction effects is a minimum. The diffraction error in electron optics is analogous to that in light optics, i.e., the radius of the aberration disk depends on l_e/γ (l_e is the electron wavelength, see below). If the electrons emitted from the small spots of the specimen exhibit a Lambert distribution, the following expression holds:

$$\gamma_0 \propto (l_e/c_s)^{1/4} \tag{2.8}$$

where for the electron wavelength we have

$$l_e \propto V^{*1/2}$$

Then δ_s is given by

$$\delta_s = l_e/\gamma_0 = C(c_s l_e^3)^{1/4} \propto \begin{cases} (c_s/V^{*3/2})^{1/4} \\ (d/V^{*3/2})^{1/4} \end{cases} \tag{2.9}$$

Here C is a factor of the order of one. For an image in the Gauss plane, $C = 0.56$ is valid. In the plane of least confusion (Scherzer, 1949), the value is somewhat smaller. The theoretical resolution in the Scherzer limit as a function of the peak field for several beam voltages is plotted in Figure 2.6.

If the optimum aperture γ_0 is used, the radius of the disk of least confusion due to chromatic aberration is given by [cor-

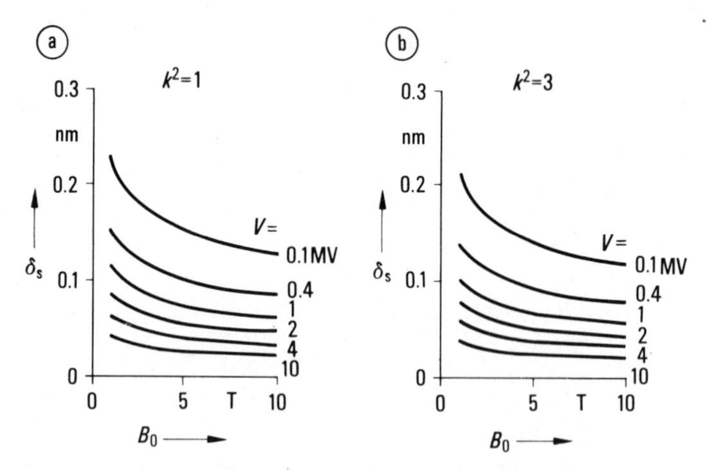

Figure 2.6. Theoretical limit of resolution δ_s (Scherzer) as a function of the peak field for different beam voltages with (a) $k^2 = 1$ and (b) $k^2 = 3$.

responding to equations (2.5) and (2.7), respectively]

$$\delta_{cV} = c_c \frac{\Delta V^*}{V^*} \gamma_0 \tag{2.10}$$

$$\delta_{cI} = c_c \frac{2\Delta I}{I} \gamma_0 \tag{2.11}$$

2.2. Rotationally Symmetric Lenses with Arbitrary Field Distribution

If there are strong deviations from a bell-shaped field, the aberration coefficients c_s and c_c and other aberration coefficients can be calculated numerically from the ray equation for rotational symmetry (Glaser, 1952):

$$\left[\frac{(2e/m_0)V^* - (e^2/m_0^2)A_\phi^2}{1 - r'^2}\right]^{1/2} \frac{\partial}{\partial z}\left[\frac{(2e/m_0)V^* - (e^2/m_0^2)A_\phi^2}{1 + r'^2}\right]^{1/2} \frac{dr}{dz} = \frac{e^2}{m_0^2}\frac{\partial}{\partial r}\frac{A_\phi^2}{r^2} \tag{2.12}$$

Here $A_\phi(r, z)$ is the azimuthal component of the vector potential, $\nabla \times \mathbf{A} = \mathbf{B}$, and $r' = dr/dz$. Equation (2.12) can be solved by using, e.g., the method described by Heritage (1973). With the iterative method of Liebmann, equation (2.12) is integrated step by step by application of a Taylor series from a point P_n to the point P_{n1}. Expansion to higher terms in r and r' and the tracing of four arbitrary independent rays through the lens permits the calculation of the third-order aberration coefficients, which include c_s.

For the case of superconducting lenses, which have not been used until now for electron micrography and similar applications, the above-described method is often too elaborate. To obtain the cardinal elements, the paraxial approximation

$$\frac{d^2r}{dz^2} = -\frac{e}{8m_0V^*} rB_z^2 = \text{const}\frac{H_z^2}{V^*}\cdot r \tag{2.13}$$

is sufficient. This is derived from equation (2.12) by expanding A_ϕ, neglecting the terms of order higher than 1 in r, and substituting

$$A_\phi = \tfrac{1}{2} r B_z$$

If numerical solutions y of the paraxial equation (2.13) are known, c_s can also be calculated from the formula (Glaser, 1952, p. 376)

$$c_s = \frac{e}{126 m_0 V^*} \int_{z_0}^{z_1} \left\{ \left[\frac{3e}{m_0 V^*} B_z^4 + 8 \left(\frac{dB_z}{dz} \right)^2 \right] y^4 - 8 B_z y^2 y'^2 \right\} dz \tag{2.14}$$

where the integration is carried out between the object plane z_0 and the image plane z_1. A similar relation holds for c_c.

2.3. Aberrations Resulting from Misalignment

We now remove the restriction of ideal alignment. Misalignment of the beam to the optical axis of the system, e.g., slanting incidence, causes a deviation from the circular symmetry in the field distribution relative to the beam axis. Defects in the field-producing elements are also sources of asymmetry. For example, if the bore in the pole pieces of a conventional lens is elliptically deformed, axial astigmatism results, i.e., the focal length differs for the two directions of the ellipse. If a difference ΔI of the objective lens current is necessary to focus the image in either the first or the second astigmatic direction, the radius δ_a of the aberration disk is given by

$$\delta_a = c_c \frac{2 \Delta I}{I} \gamma_0 \tag{2.15}$$

Image errors can also be caused by a shift of one of the pole pieces, since this produces a transverse field component superimposed on the rotationally symmetric field. This affects the deflection of the beam, so that there are additional aberrations such as anisotropic coma and anisotropic astigmatism. Sturrock (1951) and Archard (1953) treat this rather complicated problem theoretically. Their results lead to the conclusion that the deflection angle ε of the beam should be

$$\varepsilon < 5 \times 10^{-3}\,\text{rad}$$

if a resolution in the atomic region ($\approx 10^{-8}$ cm) is to be achieved.

A few more image errors are worth mentioning without entering into details:

(1) *Coma.* Slanting incidence of the beam on off-axis specimen points causes tails to appear on the image points.

(2) *Isotropic distortion.* The magnification depends on the distance between the specimen point and the z axis. Thus, off-axis object points are imaged so that a pin cushion or a barrel shape is produced if the specimen is quadratically shaped. This defect cannot be neglected for projector lenses.

(3) *Anisotropic distortion.* With increasing radial distance from the axis, the image points exhibit a growing tangential displacement. This error is caused by the spiral motion of the electrons in the magnetic field. We shall generally neglect this image rotation.

2.4. Multipole Fields for Beam Correction

Misalignment and defects in the field-producing elements which cannot be corrected mechanically can in most cases be compensated for to some degree by coil systems which generate dipole or quadrupole fields perpendicular to the lens field and thus cause deflection or cylinder lens effects.

The basic equation for the deflection of electrons moving with velocity $\mathbf{v}$ perpendicular to a uniform magnetic field $\mathbf{H}$ is given by

$$mv^2/\rho = e\mu_0 Hv \tag{2.16}$$

where ρ is the radius of curvature of the particle's trajectory and m is the relativistic mass of the particle. The deflection angle Θ is proportional to the length L of the dipole field $\mathbf{H}$ through

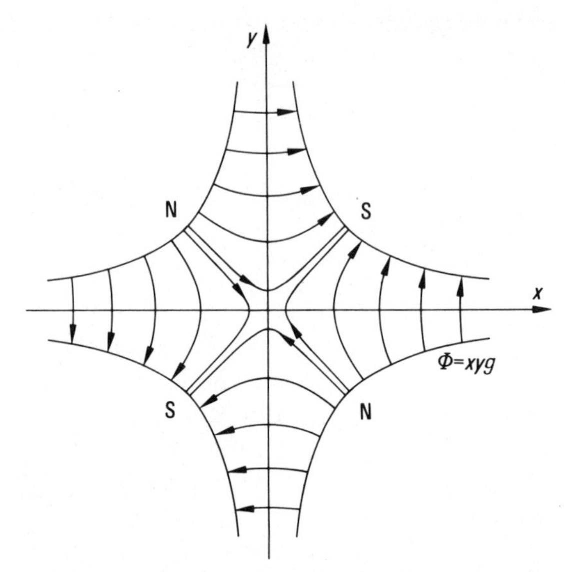

Figure 2.7. Field configuration in the cross section of a quadrupolar magnet with hyperbolic equipotential planes.

which the particle moves:

$$\Theta = \frac{Le\mu_0 H}{mv} = \left(\frac{e}{2m_0 V^*}\right)^{1/2} L\mu_0 H \tag{2.17}$$

The field distribution in the cross section of a quadrupole field is plotted in Figure 2.7. Poles of equal strength and like polarity are opposite one another. This results in a strong field gradient perpendicular to the z direction.

For a mathematical treatment it is convenient to assume that the poles, which are hyperbolically shaped, are equipotential lines and the ratio of the magnet length to the magnet aperture is greater than 1. This is in most cases a reasonable approximation. The field lines, at least in the lens center, are also hyperbolic. With the coordinate system as plotted in Figure 2.7 and with a distance $2a$ between the opposite poles, the magnetic flux density will be

$$B_y = gx$$

$$B_x = gy$$

if the potential $\Phi(\mathbf{B} = \nabla\Phi)$ is given by

$$\Phi = xyg$$

with g a constant. The potential on the pole pieces (with a distance $2a$ between opposite poles) has the form

$$g\Phi = a^2/2$$

The excitation constant β corresponding to the k value for round lenses can be derived from the equation of motion:

$$\beta^2 = \frac{K_p}{a^2}\left(\frac{2e}{m_0 V^*}\right)^{1/2} = \frac{\mu_0 NI}{a^2}\left(\frac{2e}{m_0 V^*}\right)^{1/2} \tag{2.18}$$

where $K_p = \mu_0 NI/a^2$ is the gradient of the flux density at the poles. The focal length also depends on the effective length L_1 of the field. For weak lenses ($\beta L_1 < 0.2$) such as those used in correction systems, the absolute values of the focal length in the x and y directions are equal, but there is beam convergence in one direction and divergence in the other direction:

$$f_x = -f_y = 1/\beta^2 L_1 \tag{2.19}$$

Detailed information on the electron optics of multipole fields is given by Hawkes (1966).

2.5. Image Contrast

We have discussed up to now the aberrations for specimens consisting of points emitting electrons with an isotropic angular distribution. This is a rough simplification. There are two mechanisms which produce contrast in the imaging process of a specimen, namely, scattering-absorption contrast and phase contrast. In the case of the scattering mechanism, the corpuscular theory can be applied. The elastic or inelastic scattering by transfer of energy to the electrons of the specimen atoms changes the spatial distribution of the beam amplitude (a phenomenon referred to as amplitude contrast. Suitable apertures absorb the electrons scattered at large angles (aperture contrast, Figure

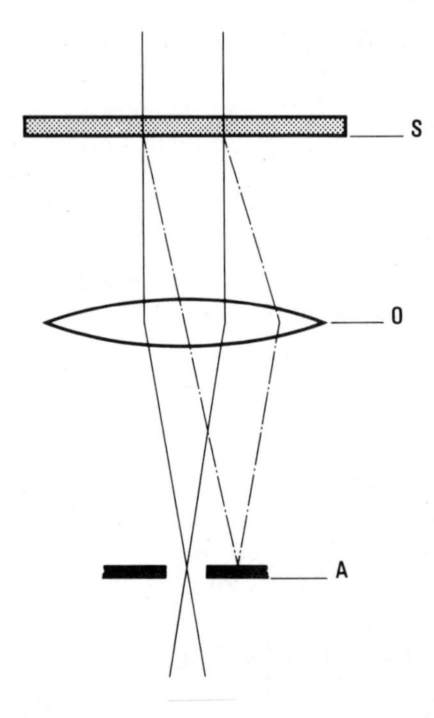

Figure 2.8. Principle of contrast due to inelastic scattering; (S) specimen; (O) objective lens; (A) aperture.

2.8). On the other hand, the wave nature of the electron must be taken into account. Phase changes of the electron waves are produced by the specimen, and the imaging process takes place by interference with undisturbed waves. Phase contrast is the most important contrast mechanism for thin specimens and single atoms, which exhibit only weak scattering. For thick specimens, the scattering-absorption mechanism also contributes to the imaging process.

Within the framework of quantum mechanics, the optical system, consisting of the objective lens and specimen, can be dealt with adequately by the electron optical transfer theory. This theory deals with the fidelity of the image produced by an optical system as a function of the spatial frequencies of the specimen. The spatial frequency spectrum is distorted by the aberration of the optical system. Using Fourier theory, one tries to express the wave function modified by the specimen in terms of the current density distribution in the image. Since the relationships in the theory of contrast are rather complicated, the following discussion of superconducting electron optical devices is in general based on the simple corpuscular theory, which considers the specimen to consist of electron-emitting points and permits the elucidation, at least semiquantitatively, of the most important aspects.

2.6. Further Sources of Error

If atomic resolution is required from an electron microscope, low-frequency electromagnetic stray fields and mechanical vibrations have to be reduced below a certain limit. The stray fields should be shielded so that the amplitude of their transverse components H_{Str} in the objective region below the specimen is such that $\mu_0 H_{\mathrm{Str}} < 0.1$ nT.

The motion of the specimen relative to the pole pieces, i.e., the drift, should be below 0.1 nm/min. A vibration-proof construction of the instrument, including a heavy foundation, is important. This is achieved without serious difficulties in the region of lower voltages, as the microscope design is then compact with all the lenses relatively small. However, the focal length $f_0 \propto d$ increases as the beam voltage is raised for a constant field H_0. As a consequence, the column has to be elongated. In addition, the lenses become larger and heavier, and the resulting larger moment of inertia and fairly low resonance frequency generally make the high-voltage microscopes sensitive to vibrations.

2.7. Fixed Beam and Scanning Mode

Two different kinds of microscopes yield high-resolution images, the fixed beam (FBEM) and the scanning transmission instrument (STEM). The optical principles of the first type are shown in Figure 2.1. The STEM is sketched in Figure 2.9. The lenses between the accelerator and specimen image the electron source on the specimen with a minimum diameter. This probe is scanned over the specimen. Because of the contrast mechanisms mentioned above (phase contrast, amplitude contrast, and aperture contrast), the electron intensity of each scan is modulated by the varying composition of the specimen. A suitable recording device allows observation of the picture on a television screen.

The resolution of the STEM depends on the electron optical parameters of the lens system and on the brightness of the electron source. It depends on the brightness because the intensity of each scan has to be so far above the noise level that the intensity

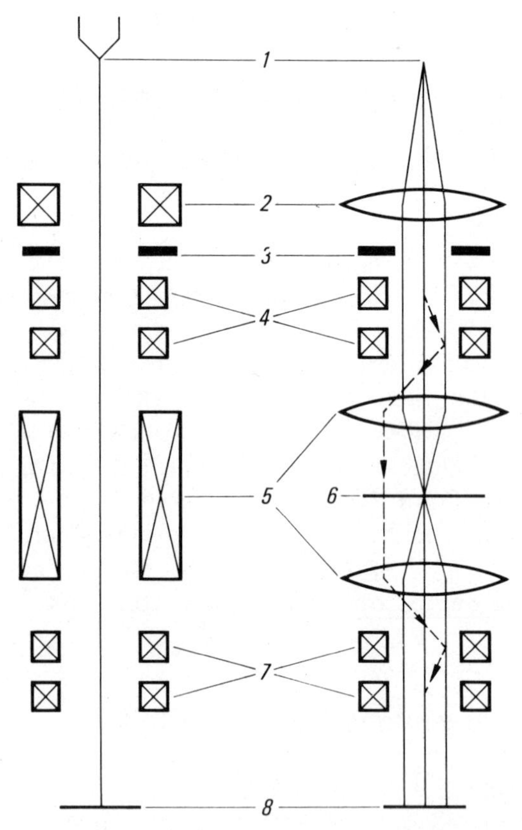

Figure 2.9. Scanning transmission electron microscope: (1) source; (2) condenser; (3) aperture; (4) two-stage deflection systems; (5) condenser objective lens; (6) specimen; (7) deflector; (8) image.

modulations are well-defined. If we neglect noise considerations, a reciprocity principle for the optical components permits a comparison of the imaging properties of STEM and FBEM. For both instruments, the aberration constants of the objective lenses* are relevant, and these lenses are arranged in front of the specimen in the STEM, behind the specimen in the FBEM. The resolution for both is given by the radius of the disk of confusion, which in a well-aligned system for the optimum aperture angle is [equations (2.9)–(2.11), (2.15)]

$$\delta = (\delta_s^2 + \delta_{cI}^2 + \delta_{cV}^2 + \delta_a^2)^{1/2} \tag{2.20}$$

The radius of the electron source which will actually be taken into consideration for the STEM can be neglected if the reduction of the lenses is chosen in a suitable way. Equation (2.20) indicates that in general the requirements for the objective lenses of STEM and FBEM are the same.

* The last demagnifying lens of the STEM is here called the objective lens for convenience.

3 | Superconducting Devices in Electron Microscopy

3.1. Advantages of Superconducting Devices

The improvements gained by superconducting devices in electron microscopy are due partly to the properties of the superconductors and partly to the low temperatures which must be maintained. However, the low temperatures give rise to the most important disadvantage, namely the necessity for cryogenic techniques.

3.1.1. Improvements Resulting from the Use of Superconducting Materials

It is readily seen that the use of superconducting materials for the lens design yields a better resolution. Chromatic aberration is reduced in superconducting lenses since they guarantee a high current stability by driving the lens coils in the persistent-current mode. The current I is only affected by flux jumps, which produce ripples that can be kept in most cases well below $\Delta I/I \approx 10^{-8}$, and by the ohmic resistance of the superconducting switch, which may cause a slow current decrease which depends on the inductivity of the device, but with in general an upper limit of $\Delta I/I = 10^{-8}$. This behavior is especially interesting if a long illumination time is required.

The current-stability problem arises not only for the objective lens coils but also for the correction system, e.g., an assembly of quadrupole and octopole lenses used to reduce the spherical and chromatic aberration (Rose, 1971). In this case, a current stability $\Delta I/I < 10^{-8}$ is required, a value which can probably only be approached using superconductors.

With superconducting lenses, one can achieve higher fields than with conventional lenses. In conventional lenses the induction is more or less limited by the saturation magnetization of the iron pole pieces, and this means that at room temperature $\mu_0 H_0 \approx 2.5$ T should not be exceeded. There is no such restriction for superconducting lenses, so one can produce very high fields ($\mu_0 H_0 \approx 6$ T) and consequently (see Figure 2.3) less than half the d value for a given voltage. The spherical and chromatic aberration coefficients are thus correspondingly reduced. This drop in the aberration coefficients occurs, however, only if the field gradient is below a certain upper limit, the value of which depends on the properties of the superconducting material used. There is one drawback: Either the bore or the gap in the lens must be large enough to leave room for the specimen. For this reason the half-width should not drop below $d \approx 2$ mm, and the enhanced peak fields are only of advantage if $V \geq 250$ kV for $k^2 = 3$ and $V > 500$ kV for $k^2 \approx 1$, respectively.

As pointed out in the last chapter, high peak fields which permit small half-widths also decrease the focal lengths. This results in a saving in size if the column is equipped with superconducting lenses. In addition, the lenses themselves are very small (Section 4.1). The scaling down becomes more spectacular as the beam voltage is increased. For 3 MV, a column consisting only of superconducting lenses should be one-third as large, but with at least the same magnification, as conventional devices (Chapter 6). Such a compact construction considerably improves the mechanical stability, as discussed previously.

A further advantage is the ease in shielding from stray fields. Magnetic stray fields with a frequency usually of 50 or 60 Hz are present in every laboratory, often with an amplitude of the order of 1 μT. Since for resolution in the atomic region, the amplitude of the stray fields near the electron beam should be $\mu_0 H < 100$ pT in a 100-kV instrument (Section 2.6), the shielding with thin single-layer mu-metal sheets, each weakening the field only by an order of magnitude or less, is a problem. Superconductors offer ideal shielding devices as long as the twofold connection is maintained.

Other superconducting elements, e.g., superconducting linear accelerators, may be of great advantage for high-voltage electron microscopy since they should reduce the size of the instrument drastically because of the high quality factor, and they might even improve the stability.

3.1.2. Improvements Resulting from the Liquid Helium Temperature in the Optical Devices

In a superconducting lens, all the walls around the specimen are at very low temperatures. Thus there is a strong pumping effect resulting in excellent vacua. This prevents the contamination of films resulting from the polymerization of organic monolayers by electron impacts and prevents the etching effect by ions of the residual gas.

Another point to be noted is that at 4 K the thermal expansion coefficient of bulk material can be as much as three orders of magnitude less than at room temperature. The low temperature of the specimen stage and its surroundings permits constructions which would have to be rejected in a room-temperature device because of high drift.

In specimens at liquid helium temperatures, Brownian motion is considerably reduced, an advantage not only for high-resolution purposes but also for diffraction studies.

Because of temperature-dependent anomalous transmission (Albrecht, 1968), the penetration coefficient in heavy atom films is greater at low temperatures. The radiation damage is significantly decreased, especially in organic materials cooled down to 4 K. At the very least, the diffusion of side chains separated from the main structure of large molecules is impeded, and so the topology is maintained. Deep-freezing the specimen and inserting it into the microscope through a cold lock prevents dehydration without using a dehydration chamber.

3.1.3. Economic Aspects

The purchase price of superconducting lenses can hardly be estimated at present. If a column with not only a superconducting objective lens but with many superconducting lenses is constructed, the superconducting device should be less expensive. A comparison of the prices of the superconducting and the normal version of a high-voltage microscope ($V \gg 1$ MV) clearly favors the former. For example, the whole superconducting machine could be installed in the basement of a normal laboratory building, whereas a special building has to be constructed for a conventional instrument. Estimates of the cost of superconducting accelerators, including liquefiers for industrial production, are not yet possible. Certainly the price of accelerators and lenses of the superconducting variety should be more or less independent of the beam voltage. For conventional instruments it increases considerably as the voltage increases.

3.2. Technical Problems

3.2.1. Construction of Cryostats

The state of the art of cryostat construction is very advanced nowadays, and normally no severe problems arise. The cryostats for superconducting accelerators and magnets in high-energy physics do not give rise to many difficulties. The difficulties with cryostats for superconducting microscope lenses are mostly connected with stability requirements. First of all, the dimensions of the connections for mechanical support should be selected to make them vibration-free. Considered from this aspect, the supporting rods should be short with a large cross section. On the other hand, the flow of heat to the liquid helium should be reduced as far as possible to avoid bubbling, which might cause motion of the specimen relative to the lens. For this reason all connections from the warmer parts of the cryostat to the helium chamber should be of high thermal resistance, i.e., the support should have low heat conductivity and appropriate geometry. Even if one tries to avoid bubbling by pumping down below the

λ point, one must keep in mind that only absorption pumps are more or less vibration-free and their pumping speed is relatively low. So in the case of pumping, good thermal insulation is still essential to maintain the He II temperature with the available pumping speed. To meet these two contradictory requirements for reducing outside and inside vibrations, one has to compromise.

Precautions have to be taken to avoid dangerous drift of parts of the lens system. One should be economical in the use of feedthroughs into the cryostat for performing mechanical motions to adjust the lenses. If feedthroughs are unavoidable, they should be made so that the low-temperature part can be decoupled when the desired position is achieved. Also, change of the low-temperature gradient in the cryostat with variation of the liquid helium level should be minimized by using, e.g., material of high thermal conductivity for the helium chamber. An essentially rotationally symmetric construction is desirable for avoiding a too strong misalignment during cooling and also for suppressing drift.

Another point which has to be considered in construction of the instrument is the prevention of contamination of the parts of the lens near the beam. The cooling should be such that the vacuum tube for the electron beam and the specimen environment are still warm while parts of the cryostat are already at low temperatures, so that cryopumping is effective. In order to keep the specimen contamination-free during the experiments, suitable apertures for heat shielding should be provided so that the aperture angle does not exceed 6° (O. Bostanjoglo, private communication). Another problem is the small space available if the superconducting lens is to replace a conventional objective in a commercial microscope.

Two different types of cryostats have been considered, namely, evaporator and bath cryostats. With one exception (Verster and Kiewiet, 1972) the latter type has been used until now for superconducting lenses since its mechanical stability is generally considered to be better. Nevertheless, the very small evaporator devices are in principle attractive, and several constructions of evaporator-type specimen stages have been published. No proposal for a microscope cryostat cooled with helium gas in the supercritical state has been published.

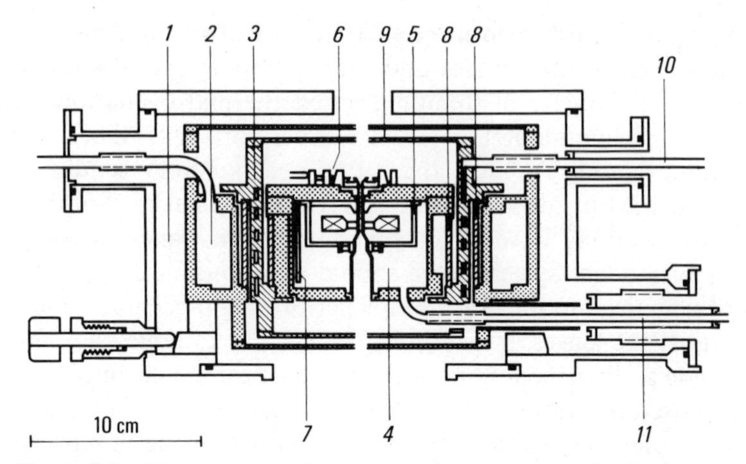

Figure 3.1. Cross section of a cryostat for a superconducting lens. Reproduced by permission of Hibino *et al.* (1973). (1) Vacuum jacket; (2) liquid nitrogen jacket; (3) helium gas heat exchanger; (4) helium chamber; (5) lens assembly; (6) specimen stage; (7) helium level indicator; (8) glass cylinder; (9) thermal shields; (10) helium exhaust; (11) helium transfer line.

An example of a cryostat fitting into a commercial microscope is shown schematically in Figure 3.1 (Hibino *et al.*, 1973). It consists of two chambers separated by a heat exchanger. The height of the outer casing is 170 mm, the diameter (without the flanges) is 260 mm. The nitrogen jacket is mounted on epoxy thermal insulators. The heat exchanger and the helium chamber are connected by glass columns. As the helium boil-off rate amounts to 1.5 liters, in practice a 3-liter liquid helium reservoir Dewar is connected directly to the relatively small helium chamber via a well-shielded transfer line, so that a lens operation time of 2 h is attained.

A cryostat with two chambers, appropriate as a housing for at least four lenses, is sketched in Figure 3.2. The cryostat was constructed to serve for lens testing. In this design, much space is available for a variety of electron optical systems and for working with superfluid helium.

The 8-liter nitrogen reservoir is connected with the outer casing by three-pointed glass-fiber rods. For maintaining stability, between the outer casing and the reservoir cover there are screw connections pressing in the vertical direction. The He I chamber (volume, 4 liters) is fastened to the nitrogen reservoir by glass-fiber tubes to which the He II chamber (volume, 4 liters) is suspended by glass-fiber rods. Adjustment screws permit alignment with the optical axis. The He II chamber is made of copper to reduce the drift when changes of the He level occur. For the transfer of motion to the objective lens, 13 double O-ring feedthroughs

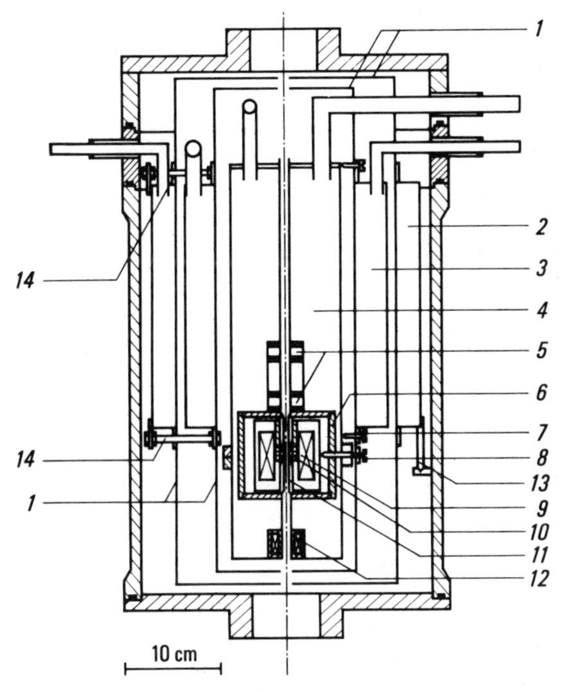

Figure 3.2. Cross section of cryostat for a superconducting lens system: (1) thermal shield; (2) liquid nitrogen jacket; (3) He I chamber; (4) He II chamber; (5) deflector; (6) superconducting shielding of the objective lens; (7) adjustment screws for displacement and tilting of the objective lens; (8) adjustment screws for coil tilting; (9) main coil of objective lens; (10) correction system; (11) shielding cylinders of objective coil; (12) intermediate lens; (13) pointed glass fiber rods; (14) glass fiber tubes.

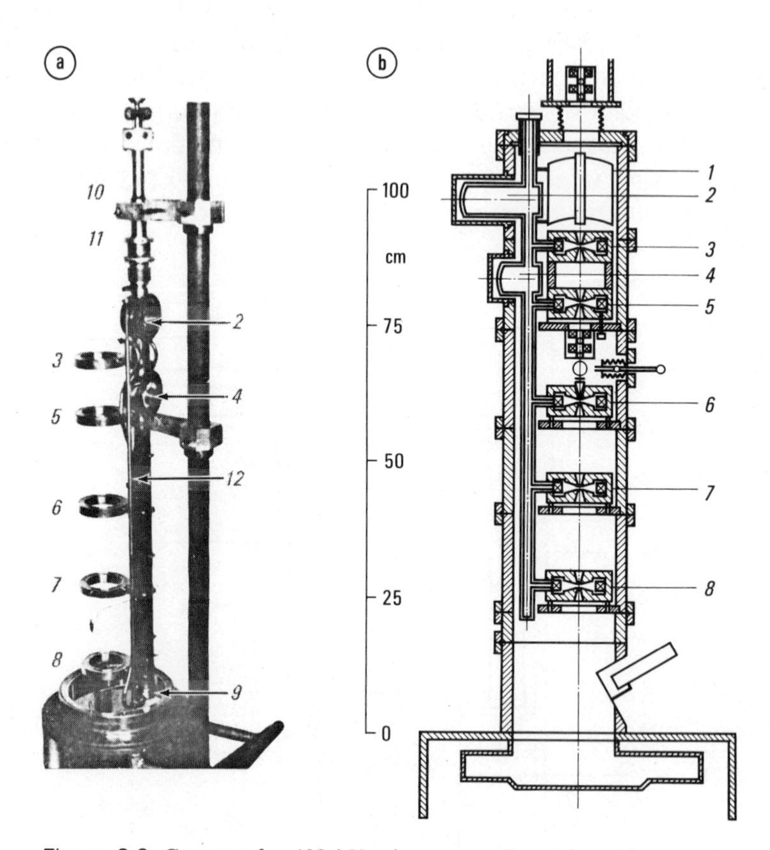

Figure 3.3. Cryostat for 400-kV microscope. Reproduced by permission of Laberrigue *et al.* (1971; 1974*b*). (a) Photograph of the cooling system; the microscope column is not yet mounted around the cryostat; the thermal shield cooled with liquid nitrogen is only attached to the vertical tube; parts of the helium reservoirs are still missing. (b) Section diagram of the column with cryostat: (1) liquid nitrogen reservoir; (2) liquid helium reservoir A; (3) condenser I; (4) liquid helium reservoir B for cooling persistent-current switches, superconducting transformers, etc.; (5) condenser II; (6) objective lens; (7) intermediate lens; (8) projector; (9) support of the cryostat tube (glass fiber); (10) neck of cryostat; (11) position for mechanical support; (12) liquid nitrogen cooled tube.

with indium seals are provided. Shielding partly prevents radiation heating. The electrical leads are inserted into the He gas exhaust and are therefore precooled before being immersed in liquid He. The evaporation rates are: N $\frac{1}{3}$ liter/h, He (He I chamber) $\frac{2}{3}$ liter/h, He (He II chamber) $\frac{1}{2}$ liter/h.

Displacement amplitudes between the outer casing and the lens system generated by mechanical vibrations are of the order of 100 nm.

Figure 3.3 shows a low-temperature setup for a microscope with superconducting lens coils (Laberrigue *et al.*, 1971). The main part of the cryostat is a tube as long as the microscope column. A helium reservoir is installed on the top of the cryostat and a second helium chamber contains the persistent-current switches. The small casings for the superconducting coils are supplied with liquid helium by branches from the main tube. All parts cooled down to 4 K are surrounded by a copper shielding kept at a temperature of 77 K by small refrigerating tubes soldered to the copper. These are connected to a liquid nitrogen reservoir. Two glass-fiber connections between each casing and the main tube prevent displacement due to electromagnetic forces. The cryostat was recently changed, the reservoirs being rearranged (Figure 5.1). The helium consumption is 3 liter/day. The problem of mechanical stability is not so important here since the iron circuits of the lenses are at room temperature.

3.2.2. Superconducting Materials

Superconductors are used in electron microscopy for several purposes:

(1) High-field superconductors are used for achieving high peak fields with large field gradients in lenses.

(2) Superconductors are used for microwave cavities in linear accelerators since they exhibit a high quality factor.

(3) Superconductors are used for shielding stray magnetic fields of low frequencies as substitutes for mu-metal casings in conventional microscopes.

In case (3) elemental or alloy superconductors with a critical temperature far above 4 K and appropriate mechanical properties

Table 3.1. The Superconducting Elements in the Periodic System[a]

H																	He
Li	Be 0.03											B	C	N	O	F	Ne
Na	Mg											Al 1.19 9.9	Si 7	P 6	S	Cl	A
K	Ca	Sc	Ti 0.39	V 5.3 131	Cr	Mn	Fe	Co	Ni	Cu	Zn 0.9 5.3	Ga 1.09 5.1	Ge 5.4	As 0.5	Se 7	Br	Kr
Rb	Sr	Y 2.7	Zr 0.55 4.7	Nb 9.2 194	Mo 0.92	Tc 7.8	Ru 0.5 6.6	Rh	Pd	Ag	Cd 0.55	In 3.4 29.3	Sn 3.7 30.9	Sb 3.6	Te 4.5	I	Xc
Cs 1.5	Ba 5.1	La 5.9 160	Hf 0.16	Ta 4.48 8.3	W 0.01	Re 1.7 20.1	Os 0.65 8.2	Ir 0.14	Pt	Au	Hg 4.15 41.1	Tl 2.39 17.1	Pb 7.2 80.3	Bi 8.5	Po	At	Rn
Fr	Ra	Ac	Th 1.37 16.2	Pa 1.3	U 0.2												

RARE EARTHS						
Ce 1.7	Pr	Nd	Pm	Sm	Eu	Gd
To	Dy	Ho	Er	Tm	Yb	Lu 0.7

[a] The elements outlined by solid lines are superconductors at atmospheric pressure; the dashed lines indicate high-pressure phases. In each box, the upper number gives the critical temperature in K, the lower number gives $\mu_0 H_c$ or $\mu_0 H_{c2}$ in mT, where H_c and H_{c2} are the critical fields.

are usually suitable. Table 3.1—a periodic table with critical temperatures and critical fields H_c, H_{c2}, respectively, marked on it—indicates that, for example, lead or niobium can be used, i.e., the type of superconductor is unimportant. (Appendix B gives a short survey of the different types of superconductors, in particular the imaging of their domain structures in electron microscopes.)

The following section discusses the tougher technical requirements for superconducting magnetic-field-generating and accelerator devices. The quality of superconducting lenses is determined chiefly by the properties of the superconducting material used for the magnetic circuit. For high peak fields and field gradients, the basic requirements are large effective critical current densities. In Figure 3.4 the critical current density as a function of the applied external flux density is plotted for commercially available high-field superconductors and those which may become interesting in the near future. It is obvious that Nb_3Sn with the addition of Cu, V_3Ga, Nb_3Ge, and similar compounds with β tungsten structures (A15 phases) are most suitable between 5 T and 8 T.

However, for coil applications one has to consider that A15 compounds are brittle and require a considerable amount of normal-conducting embedding material for mechanical and electrical stabilization. Even in diffused multifilamentary wires the overall critical current density i_c in the wire scarcely exceeds 2×10^5 A/cm^2 at 5 T. The bending radius is normally limited to about 5 mm without degradation. Even this geometry can only be reached if the coils are wound with unreacted wire, and this gives rise to insulation problems. Hence it is difficult to obtain very high field gradients on the axes of superconducting coils.

In this respect, it is preferable to replace the coils by superconducting rings with trapped flux. A ring is excited by applying an external field, which induces shielding currents. These currents, once induced, prevent external fields from penetrating into the interior of the ring. The highest field which can be shielded by the ring is equal to the maximum field used in trapping flux in the ring. For a hollow cylinder in an external field parallel to the axis (Figure 3.5), the shielding values can be calculated if the whole sample is assumed to be in the critical state (i.e., if

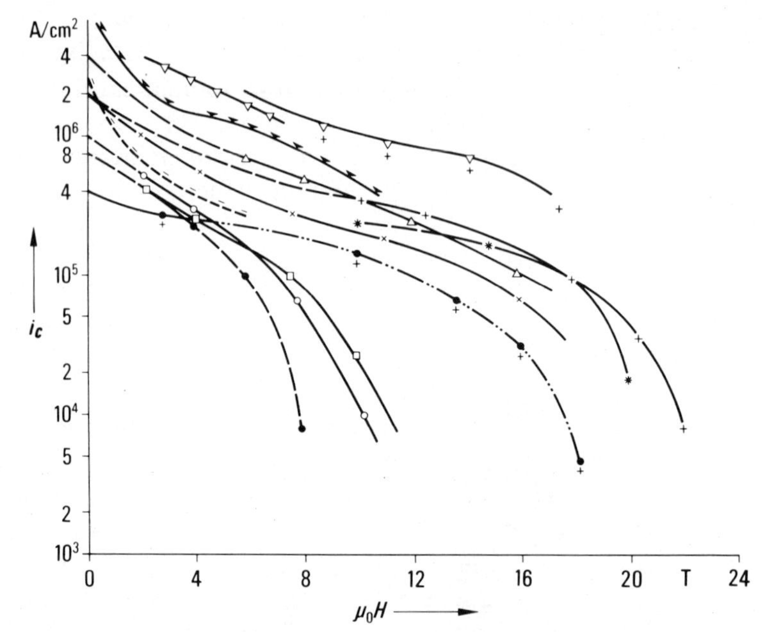

Figure 3.4. Short-sample critical current densities for some high-field superconductors: (▽) Nb_3Sn + C (Ziegler *et al.*, 1971); (▽ over +) V_3Ga (Howe and Weinmann, 1974); (▼) Nb_3Ge (Tarutani *et al.*, 1974); (△) Nb_3Sn + Cu (Caslaw, 1971); (+) NbN (Deis *et al.*, 1969); (×) Nb_3Sn (Cody and Cullen, 1964); V_3Ga:(−)(Wilhelm, 1972), (∗)(Tachikawa and Iwasa, 1970); (□) Nb/50 Ti (Vacuum Schmelze Hanau); (○) Nb/65 Ti (Vacuum Schmelze Hanau); (●) Nb/33 Zr (Vacuum Schmelze Hanau); (● over +) $V_2Hf_{0.5}Zr_{0.5}$ (Inoue *et al.*, 1971).

the shielding currents are induced in the whole cross section) by integrating the quenching curves of Figure 3.4. For Nb_3Sn with carbon additions the theoretical shielding capability for a wall thickness $w = 1$ mm would be (Dietrich *et al.*, 1972*b*)

$$\mu_0 H_{s1} \approx 13.5\ \mathrm{T}$$

This cannot be achieved in practice due to flux jumps and other instabilities. The experimental values of the shielding capability $\mu_0 H_{s1}$ can be obtained from the tube diagram (Figure 3.5). If the following relation (Anderson, 1962; Kim *et al.*, 1963) is valid

$$i_c(B)\,B = \mathrm{const} \tag{3.1}$$

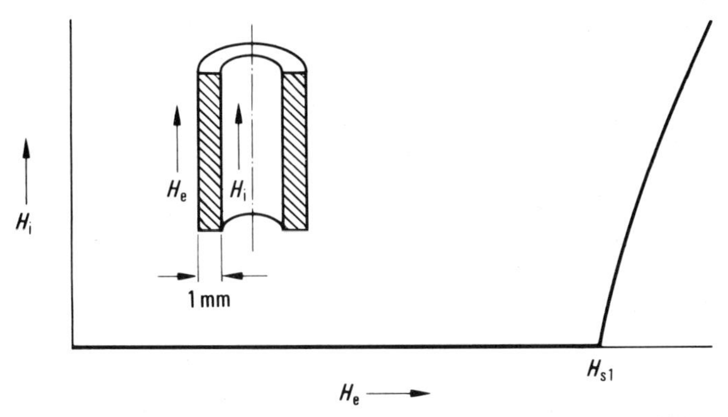

Figure 3.5. Tube diagram for obtaining the shielding capability. H_{s1} is the shielding field for a 1-mm thickness, H_e the external field, and H_i the internal field.

which is approximately true for the case of A15 phases, the shielding capability $\mu_0 H_{sw}$ for wall thickness w, as well as the critical current density $i_c(H_{sw})$, can be easily derived since

$$H_{sw} \propto w^{1/2}$$

and

$$i_c(H_{sw}) = H_{sw}/0.2w$$

where i_c is in A/cm^2, H_{sw} in A/cm, and w in cm. The data of Table 3.2 indicate that the effective critical current density of specially

Table 3.2. Shielding Values for Some Materials[a]

Material	w, mm	$\mu_0 H_{sw}$, T	$\mu_0 H_{s1}$ (eff), T	Reference
Nb-10/Ti-40/Zr	2.0	1.9	1.4	Kawabe *et al.* (1969)
V_3(GaAl)	1.5	1.55	1.25	Müller and Voigt (1970)
Nb_3Sn, vapor deposited	0.015	0.65	6.9	Hanak (1964)
Nb_3Sn stacks of Cu-plated tape	5	4.2	1.9	Firth *et al.* (1970)
Nb–Sn sinter material hammer-dressed with Cu additions	1	8.0	8	Dietrich (1974)

[a] w is the wall thickness; H_{sw} the shielding field for a given w (mm), and $\mu_0 H_{s1}$ (eff) the shielding capability for the combination of superconducting and normal material.

treated Nb–Sn sinter material with Cu additions amounts to nearly 3×10^5 A/cm^2 at 8 T. As a consequence, very high field gradients are feasible.

For coils and bulk superconductors good electrical stability is a condition *sine qua non* in electron optics. Electrical stability means on the one hand that the excitation can take place at a high rate without quenching, and on the other hand, that no spontaneous flux jumps occur in the material. The stability of a hammer-dressed bulk cylinder as a function of the field exciting rate $\dot{H}_e$ is shown in Figure 3.6. The jumping field H_j is 2% below the maximum shielding field if $\mu_0 \dot{H}_e = 0.1$ T/min. For this low exciting rate, degradation can be neglected. Coils of relatively stable material with an effective current density an order of magnitude smaller may be excited with $\mu_0 \dot{H}_e \approx 10$ T/min and higher without quenching.

In order to be unaffected by the noise of the power supplies, the coils are driven in the persistent-current mode. The principle is shown in Figure 3.7. The two free ends of the coil are welded or soldered to a superconducting thick wire for support. During the coil excitation, part of the superconductor in the short-

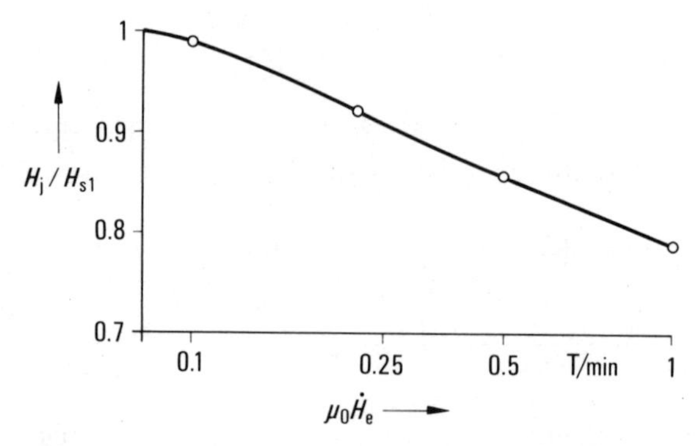

Figure 3.6. Stability of a sample of Nb–Sn sinter material with Cu additions (67.7 Nb–28.8 Sn–3.5 Cu). Annealing temperature 750°C, annealing time 2 h. The excitation rate $\mu_0 \dot{H}_e = 0.1$ T/min given in the diagram is applied when $H_j/H_{s1} \geq 0.8$. Up to this ratio, $\mu_0 \dot{H}_e = 0.25$ T/min is feasible. H_j is the jumping field H_e the external field, and H_{s1} the shielding field for a wall thickness of 1 mm.

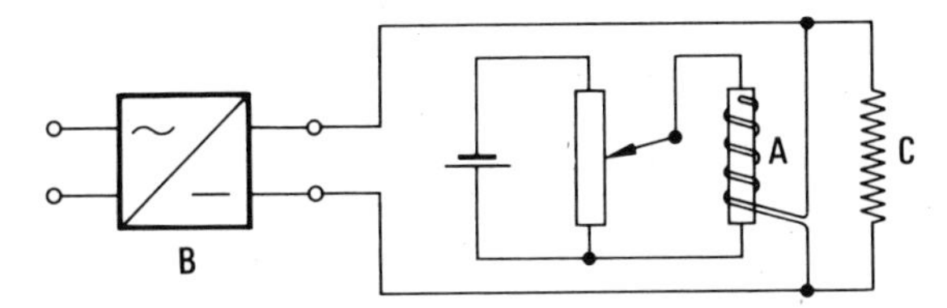

Figure 3. 7. Coil circuit with persistent-current switch: (A) heater; (B) current generator; (C) coil.

circuit in Figure 3.7 is transferred into the normal state by heating. As soon as the desired current strength is obtained, the heater is switched off and the superconducting circuit is closed. The inductivity of the short-circuit should be several orders of magnitude smaller than the coil inductivity to avoid a decrease of the current as a consequence of the switching process.

Since current ripple and drift should be below $\Delta I/I = 10^{-8}$/min, the problem of making the contacts between the superconductors reliable and with negligible resistance arises in some cases. This is a particularly difficult problem in the case of minicoils with small inductivities made from multicore wire. However, correction coils with an outer diameter of 8 mm and a bore of 4 mm made of 50-μm Nb–Ti single-core wire can exhibit a drift $\Delta I/I < 10^{-6}$/min. If focusing by the lens current is desired, it is better to use transformers for the current variation rather than open the persistent-current switch. The focusing could also be done by using shim coils.

The requirements of rotational symmetry for iron-circuit superconducting lenses are identical with those for conventional lenses. However, in lenses which consist of simple coils or strongly oversaturated ferromagnetic systems in which the coil field is overwhelming, deviations from the circular shape of the coils are critical. The main axes of the solenoids should not differ by more than a few micrometers, i.e., the coil form must be machined very carefully.

The rotationally symmetric geometry may also be disturbed by forces acting on the coils. To prevent loose windings, a certain stress must be applied to the superconducting wire during coil fabrication. The thermal expansion of the materials used in the coil, e.g., superconductors and the matrix in composite superconductors, differs in most cases, and thus forces due to thermal contraction occur during cooling. Circumferential stresses caused by the Lorentz forces **i** · **B**, and which can be neglected in normal-conducting coils with low current densities, are important in superconducting designs.

Winding steps might also be dangerous. This problem can be avoided by using tape as shown by Mulvey and Newman (1972) for water-cooled iron-free coils. A problem which arises on replacing the copper by a superconductor is the high current in the ribbon, since this can produce a strong field distortion at the beginning and the end of the coil. However, according to Mulvey and Newman, a wedge cut into an otherwise circular bore which introduces a similar distortion should not much affect the image aberrations.

If bulk superconductors replace the pole pieces, machining of the material with tolerances in the micrometer region is decisive, and the penetration of the field into the superconductors should be uniform, i.e., the pinning centers responsible for the current density should be homogeneously distributed. A rough indication of the "homogeneity" of a superconductor is given by a metallographic micrograph. In Figure 3.8 two anodically oxidized Nb–Sn sinter-material samples are compared; the one on the left was made of coarse Nb powder (grain size 40 μm), hydrostatically pressed and annealed for 2 h at 900°C, while the one on the right side was made of fine-grain Nb powder (average diameter 8 μm) mixed with Sn-rich Nb phases and Sn powder pressed in a Ni tube, hammer-dressed, and sintered at 750°C for 2 h. When used in a lens in place of Fe pole pieces, the material with the more uniform microstructure exhibited only small deviations from the desired rotational symmetry.

If microwave accelerators are to be used in electron microscopy, only superconducting devices can be chosen. More details on linear accelerators will be given in Chapter 6. We will discuss here the material for resonators. In contrast to the superconductors in magnetic devices, where the volume properties of the superconductors are responsible for their behavior, only the surface properties of superconductors are relevant in high-frequency

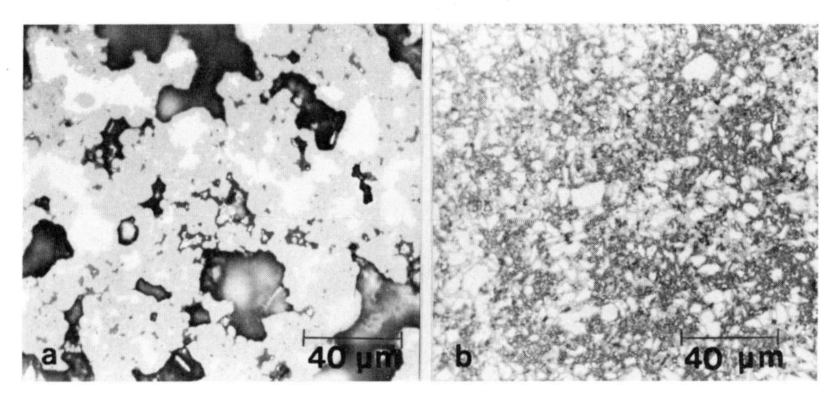

Figure 3.8. Anodically oxidized specimens of Nb–Sn sinter material [Dietrich *et al.* (1972*b*)]. The light areas are Nb, the medium grey are Nb_3Sn, and the dark grey and black are other Nb–Sn phases and voids. (a) Material made of coarse-grain Nb powder. Sinter temperature 900°C, sinter time 2 h, $\mu_0 H_{s1} \approx 2.5$ T/mm. (b) Material made of fine-grain Nb powder with Cu additions. Sinter temperature 750°C, sinter time 2 h; $\mu_0 H_{s1} \approx 7.5$ T/mm.

applications. The surface impedance of a superconductor is given by

$$R \propto R_0 \omega^2 \exp(-\Delta/kT) \tag{3.2}$$

where R_0 is the residual surface resistance, ω the frequency, Δ the energy gap, and k Boltzmann's constant.

Experimental values of the temperature-dependent surface resistance of a Pb cavity are shown in Figure 3.9. For temperatures below the λ point and frequencies in the GHz region, the impedance approaches saturation as determined by R_0. It is evident that the energy gap of the superconductor should be relatively large and the residual surface resistance R_0 be as small as possible for obtaining a low impedance. Since the energy is approximately proportional to the critical temperature, high-temperature superconductors are chosen, e.g., Nb and Pb; Nb_3Sn ($T_c \approx 18$ K) and NbN ($T_c \approx 15$ K) have also been proposed. The residual surface resistance of Nb can be lowered to 10^{-9} Ω, a value which is to be compared with 10^{-4} Ω in the case of Cu at room temperature. This extremely small value can be achieved by special surface treatment of the resonators. After annealing in an ultrahigh vacuum, the surfaces are oxipolished for smoothing and to prepare

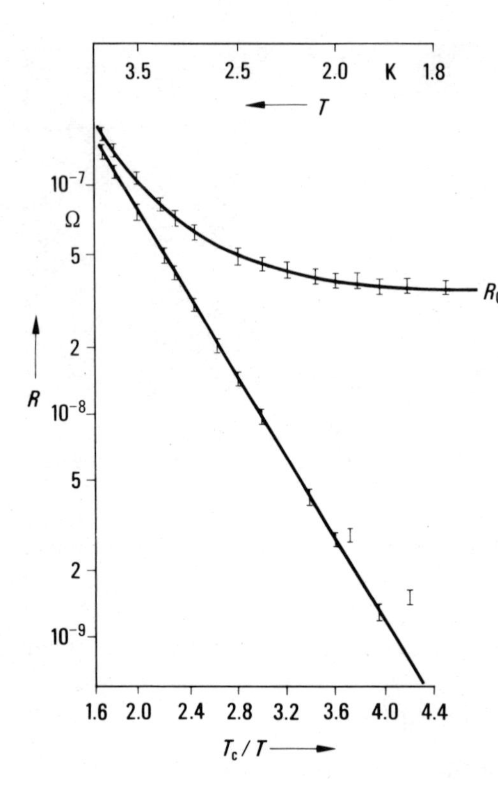

Figure 3.9. Surface resistance of a Pb cavity at 380 MHz as a function of the reciprocal temperature. The straight line corresponds to relation (3.2). It is obtained by subtracting the constant value R_0, the residual surface resistance, from the measured points of the upper curve. The critical temperature T_c of Pb is 7.3K.

them for deposition are covered with a protective layer by electropolishing (Diepers *et al.*, 1973). Cavities are normally characterized by the quality factor Q, which is defined as the ratio of a geometric factor to the impedance. If the quality factor amounts to $Q > 10^9$, a field strength of 4 MV/m is obtainable. Values of Q as high as $Q = 10^{10}$ have been reached frequently. Thus a 3-MV accelerator of about 1-m length, as proposed in Chapter 7, is not utopic.

3.2.3. Normal Materials

For lenses with ferromagnetic circuits, the low-temperature properties of magnetic materials should be known. They have not yet been studied extensively for iron and those of its alloys which are used for the ferromagnetic circuits in the room-temperature lenses. The stability with respect to Barkhausen jumps is

Table 3.3. Temperature Dependence of the Relative Permeability of mu-Metal and Cryoperm[a]

Material	300 K		75 K		4 K	
	μ_{ri}	$\mu_{r\max}$	μ_{ri}	$\mu_{r\max}$	μ_{ri}	$\mu_{r\max}$
mu-Metal	99,000	230,000	12,000	26,000	9,000	19,000
Cryoperm	32,000	260,000	36,000	250,000	31,000	235,000

[a] Data supplied from the Vacuumschmelze Hanau. μ_{ri} is the initial permeability and $\mu_{r\max}$ the maximum permeability.

not quite established. This is also true for the soft magnetic alloy Co–Fe. This alloy has the highest known saturation magnetization at room temperature and therefore is frequently used for pole pieces in room-temperature devices.

As mentioned previously, electromagnetic fields are easily screened in cryostats by superconductors. However, as constant fields are trapped in superconductors, shielding from such fields before cooling is sometimes necessary. This can be achieved by magnetic materials with high permeability.

There is often quite a difference between room-temperature and helium-temperature behavior in these materials. This is indicated by the magnetic data for the two soft magnetic alloys listed in Table 3.3. Cryoperm is preferable to mu-metal at low temperatures as the permeability is temperature-independent.

Table 3.4. Ferromagnetic Materials Suitable for Pole Pieces in Superconducting Lenses[a]

Material	Curie temperature, K	Saturated magnetization, T	μ_{ri}	$\mu_{r\max}$	Reference
Ho	20	3.35	5	12	
Ho monocrystal (*b* axis)	—	3.75	1.5	500	Bonjour (1973)
Dy	85	3.05	3	8	
Fe	1040	1.6	1300	3300	
Co–Fe 50 alloy (Vacoflux)	1220	2.35	440	600	

[a] μ_{ri} is the initial permeability at 4 K, and $\mu_{r\max}$ the maximum permeability at 4 K.

Table 3.5. Mechanical and Thermal Properties of Some Cryotechnical Materials[a]

Material	T_e, N/mm^2			E_{02}, N/mm^2			$\frac{1}{L}\frac{dL}{dT}$, 10^{-5}K^{-1}		$\frac{\Delta L}{L}$,	κ, W/mm K	
	300 K	90 K	4 K	300 K	90 K	4 K	300 K	4 K	10^{-3}	300 K	4 K
A/S/steel 303 Fe–C–Mn, Si, Co, Ni, Mo	750–1000	—	1800–2000	—	—	—	—	—	—	—	—
A/S/steel 304 stainless steel	750–1800	—	1600–1800	300	—	900	1.6	—	3	27	3.5
Electrolyte Cu	250	400	—	60	80	—	1.7	0.002	3.4	10^3	2.10^3
Special brass (56 Cu/1.4 Mn/1.2 Pb/1 Fe)	500	700	—	176	200	—	—	—	—	—	—
Glass filament epoxy											
along the filament (‖)	300	—	710	—	—	—	—	—	3.5	0.8	0.1
across the filament (⊥)	—	—	—	—	—	—	—	—	7	1.3	0.2

[a] T_e is the tensile strength, E_{02} the yield strength, $(1/L)(dL/dt)$ the thermal expansion coefficient, $\Delta L/L = (L_{293} - L_4)/L_{293}$ is the thermal contraction, and κ is the thermal conductivity. The data are taken partly from *Cryogenic Materials Data Handbooks*, 1962–1970 (U.S. Department of Commerce, Clearinghouse for Federal Scientific and Technical Information), partly from Conte (1970).

At helium temperature, holmium and dysprosium, the rare earths with low Curie temperatures and the highest known saturation magnetizations, are also interesting for use as pole pieces. Some of their magnetic properties are compared with those of iron and Vacoflux in Table 3.4. The relative permeability of the polycrystalline rare earth samples is low. As a consequence, coils with more ampere turns are required than in the case of iron alloys. Holmium and dysprosium are brittle and not easy to handle. For example, their ultimate tensile strength amounts to 230 N/mm^2. Practical experience in the field of electron microscopy is still limited for these materials.

All the normal metals and insulating materials used in cryostats and in the lenses have to fulfill certain requirements in tensile strength, brittleness, and other mechanical properties. Further, the thermal expansion coefficients of the materials must be coordinated. The thermal conductivity is also important in many cases. Unfortunately, the necessary low-temperature data of many commercial materials are often not available, in particular those of the different varieties of steel. Sample data for some materials used in the superconducting shielding lens (Section 4.1.5) are given in Table 3.5.

A further important point is the magnetic behavior of steel and German silver. It is essential that the support materials within a lens near the beam exhibit only weak para- or diamagnetism. Several steel and German silver varieties are known to become more ferromagnetic at low temperatures. For example, it is reported that the relative permeability of some stainless steel samples was $\mu_r = 1.2$ down to 80 K. Below that temperature a ferromagnetic transition took place. The steel alloys given in Table 3.5 were tested at 4 K with the result $\mu_r = 1.01$ at $T = 4$ K.

4 | Lens Design and Testing

4.1. Lens Design and Field Distribution

4.1.1. General Remarks

Different types of superconducting rotationally symmetric lenses for which field distributions have been obtained are described in the literature. The axial field measurements were usually carried out with low-temperature Hall probes, magnetoresistance probes, or probe coils. In one case the field distribution in the whole lens was determined from a model in an electrolytic tank (Dietrich and Weyl, 1968). Yet the normal procedure for obtaining the field distribution is to calculate it with a computer program. For simple coils, the Biot–Savart method is appropriate. In the case of more complex boundary conditions, the Laplace equation has to be solved. Using **A** for the vector potential, we have

$$\mathbf{B} = \nabla \times \mathbf{A}$$

$$\nabla \times \mathbf{H} = \mathbf{i}$$

and in cylindrical coordinates the following relation holds:

$$\frac{2\partial A_\phi^2}{\partial r^2} + \frac{1}{r}\frac{\partial A_\phi}{\partial r} + \frac{\partial^2 A_\phi}{\partial z^2} = \mu_0 i + \frac{A_\phi}{r^2} \tag{4.1}$$

since in a rotationally symmetric system $A_r = A_z = 0$.

The methods used to date for the numerical solution of equation (4.1) for superconducting lenses are the differential techniques proposed by Liebmann (1949), for which examples are given by Dietrich *et al.* (1972*a*), and the more elaborate finite-element method (Munroe, 1973). If deviations from the rotational symmetry are taken into consideration, the Green's function method might be helpful. Results for this case are not yet available. Numerical calculations of the field distribution in magnets with partly iron circuits are outlined by Brechna (1973).

4.1.2. Coil Lens

As mentioned in the historical survey, the first electron microscope image achieved with a superconducting device was made with a simple coil lens (Fernández-Móran, 1965). The field along the coil axis is calculated easily using (Montgomery, 1969)

$$H_z = H_0 \frac{F\left(\frac{1+z}{r}\right) + F\left(\frac{1-z}{r}\right)}{2F(1/r)}$$

$$H_0 = i\lambda r F(1/r) \times 10^3$$

$$F(u) = u[\sinh^{-1}(R/ur) - \sinh^{-1}(1/u)] \tag{4.2}$$

Here i is the current density in A/mm^2, H_z the field along the axis in A/m, λ the packing factor, r the bore radius of the coil in mm, R the outer radius of the coil in mm, $2l$ the length of the coil in mm, and $u = (l + z)/r$ or $(l - z)/r$.

Figure 4.1 shows the lines of force and the field along the z axis for a coil with $r/R = 1/4.7$. More favorable results for electron microscope purposes are obtained with pancake coils as shown in Figure 4.2 (Génotel *et al.*, 1967). In Figure 4.3 (Laberrigue and Severin, 1967) the peak field as a function of d is plotted for several coil geometries. Such geometries are in principle suitable for lens designs based on commercially available

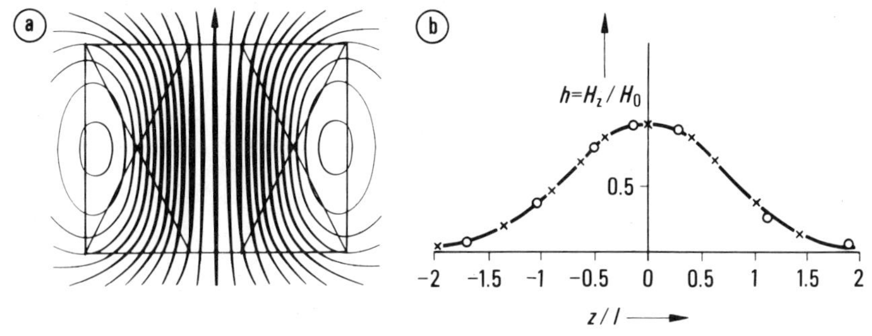

Figure 4.1. Field distribution in a solenoid. (a) Calculated distribution. The width of the field lines is proportional to $|\mathbf{H}|$). (b) Axial field: (○) calculated values; (×) experimental values.

wires which do not exhibit degradation at a small radius of curvature.

If we assume a solenoid with $r = 2$ mm, $R = 18$ mm, and $l = 2$ mm, and an effective critical current density $i_c = 10^5$ A/cm^2, then a maximum induction $\mu_0 H_0 = 5$ T and $d = 3.5$ mm can be expected. According to relation (2.2) this would correspond to $V \approx 2$ MV for a $k^2 = 1$ lens. But there are problems connected

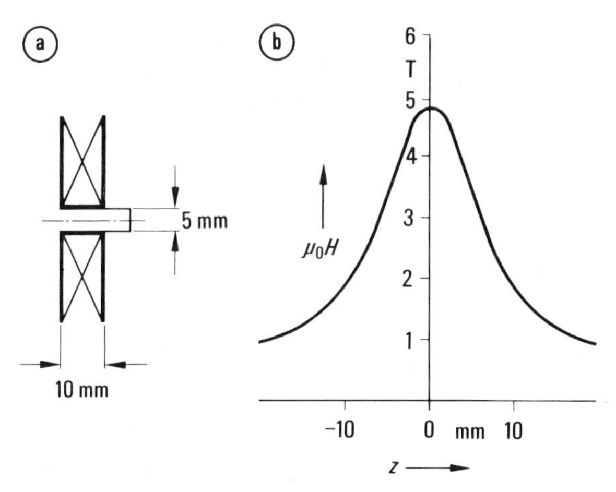

Figure 4.2. Pancake coil (a) and its field distribution (b). Reproduced by permission of Génotel *et al.* (1967).

with the windings of coils, especially for such small devices as were discussed in Section 3.2.

Improvement may be gained by adding coils to the main coil but with reverse current. This reduces the half-width but does not lower the peak field to the same degree. Merli (1970) designed a system consisting of a main coil and three auxiliary coils on one side (Figure 4.4a). The axial field distribution is strongly asymmetrical for this geometry (Figure 4.4b). The peak field is shifted relative to the median plane of the main field and relative to the principal and the focal planes as well. The advantage is a relatively large distance h between the main coil and the specimen location that makes it easy to construct a suitable specimen holder. The aberration coefficients calculated by numerical integration for the field distribution plotted in Figure 4.4b are small considering the h value of 3.7 mm.

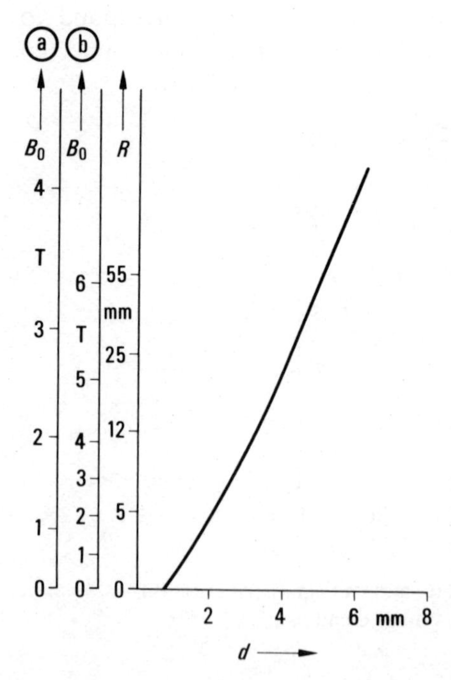

Figure 4.3. Peak induction B_0 as a function of d for solenoids with varying outer radius R, with r = 2 mm, l = 2 mm. Scale a is for Supercon Nb–Zr wire with $i_{eff} = 5 \times 10^4$ A cm^{-2} at 2.5 T. Scale b is for Nb_3Sn ribbon with $i_{eff} = 10^5$ A cm^{-2} at 5 T.

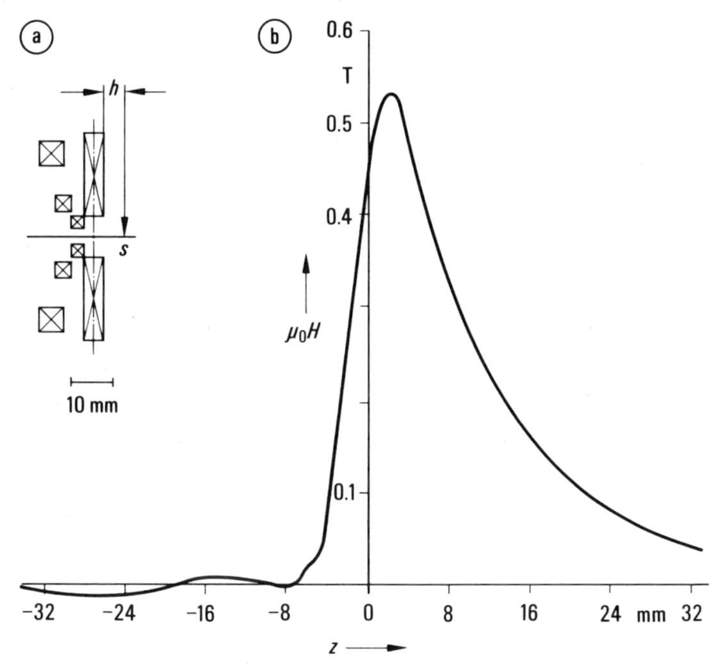

Figure 4.4. Field distribution (b) for the system of superconducting coils shown in (a). Reproduced by permission of Merli (1970). Current density $i_{\text{eff}} = 2 \times 10^4$ A cm^{-2}; beam voltage $V = 100$ kV; $c_s = 2.7$ mm; $c_c = 3.4$ mm; the distance h from the main coil to the specimens is 3.7 mm.

4.1.3. Ring Lens

A more elegant method than the one just discussed is to use the trapped flux in a superconducting ring. In principle the desired field H_m is produced while the ring is in the normal state. The ring is cooled below T_c in this field, and the field then switched off (Figure 4.5). Because of the induction law, a decrease of the external field excites currents in the cylinder, and these do not diminish in the superconducting material, since the superconductor behaves like an ideal conductor.

Figure 4.6a shows the field distribution in and around the ring when the current fills the whole cross section. Figure 4.6b gives the field along the z axis, together with experimental values.

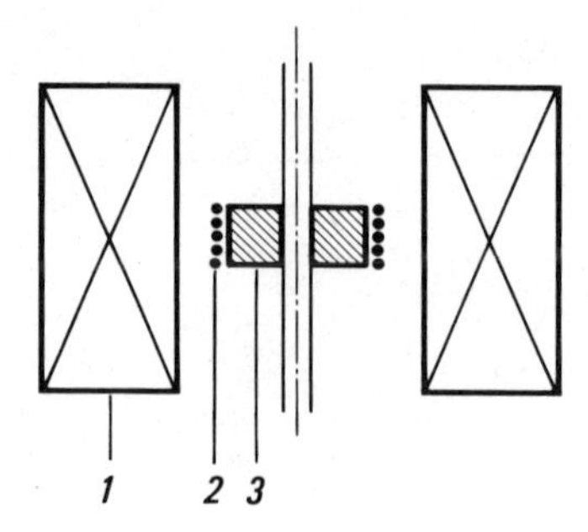

Figure 4.5. Ring lens: (1) superconducting coil; (2) heater; (3) superconducting ring.

At first glance the field distributions in Figures 4.2 and 4.6 appear similar. However, they differ in that the current density in the superconducting ring varies with the local field, while there is a constant current density in the coil of Figure 4.2.

For numerical calculations we assume as a first approach that a homogeneous field H_m is trapped in the ring (Dietrich *et al.*, 1972*a*). For this model we have

$$H_m = \frac{1}{\mu_0}\left(\frac{\partial A_\phi}{\partial r} + \frac{A_\phi}{r}\right) = \text{const} \tag{4.3}$$

The solution in the ring takes the form

$$\frac{1}{\mu_0} A_\phi = \frac{H_m}{2} r + \frac{C}{r}$$

where C is an integration constant; since the integration boundary contains the axis, $C = 0$. This relation can be used as a boundary condition on the ring if there are only surface currents. If the current density exceeds a critical value $i_c(H)$ given by the Kim–Anderson relation (3.1), the current-carrying zone penetrates into the superconductor. The current density in this region, which is obtained from equation (4.1), is calculated by iteration, using the previously obtained value of the local field. A second condition is that the local field cannot exceed H_m. The iterations are continued until relation (3.1) is fulfilled in each element of the superconductor. [Each $i_c(H)$ relation, even if not expressed by an analytic function, can be used for these calculations.]

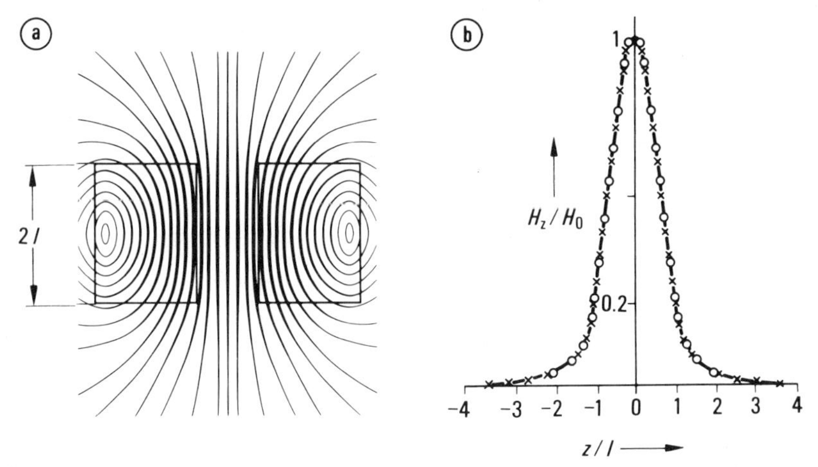

Figure 4.6. Field distribution in a ring lens (lens B, Table 4.1). (a) Calculated distribution (the width of the field lines is proportional to $|\mathbf{H}|$). (b) Axial field: (○) calculated values; (×) experimental values. $2l$ is the length of the ring.

The data for three ring lenses studied by different authors are listed in Table 4.1. Ring 1 is made of a stack of Nb_3Sn disks with a ratio of normal to superconducting material of 3:1 and

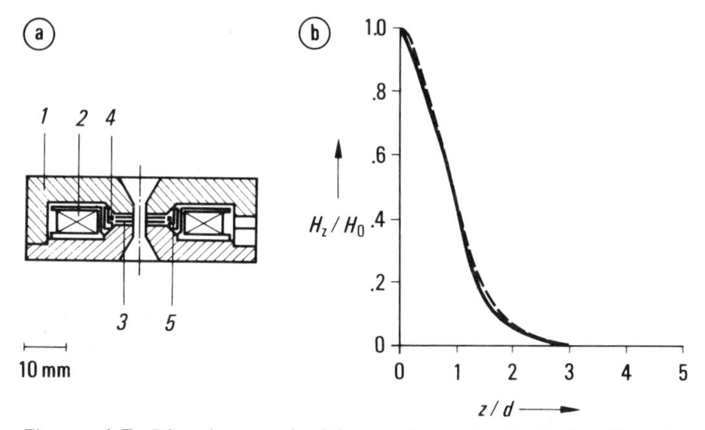

Figure 4.7. Ring lens excited by an iron circuit device (lens A, Table 4.1). Reproduced by permission of Hibino *et al.* (1973). (a) Lens assembly: (1) ferromagnetic container; (2) excitation coil; (3) stack of Nb_3Sn disks; (4) heater; (5) Teflon insulator. (b) Field distribution. The solid curve is the experimental field, the dotted line the Gaussian field.

Table 4.1. Data for Ring Lenses[a]

Lens	Ring dimensions						Bell-shaped field approximation					Numerically calculated				Reference
	R, mm	r, mm	l, mm	$\mu_0 H_0$, T	$2d$, mm	V, kV	k^2	f_0, mm	$z(f_0)$, mm	c_s, mm	c_c, mm	f_0, mm	$z(f_0)$, mm	c_s, mm	c_c, mm	
A	6.3	1.5	1.3	2.1	3.5	100	3	1.75	0	0.5	1.0	1.2	0.1	0.7	0.9	Hibino *et al.* (1973)
				2.1	3.5	250	1	2.6	1.2	1.2	1.5	2.1	−1.4	1.5	1.6	
B	4.2	1.0	2.1	7.0	5.8	1300	3	2.9	0	0.9	1.8	2.1	−0.2	1.1	1.5	Dietrich *et al.* (1972*a*)
				7.0	5.8	2000	1.5	3.2	−1.3	1.4	2.1	2.7	−1.4	1.7	1.9	
				4.7	6.8	1000	2.8	3.4	−0.1	1.0	2.0	2.4	−0.2	1.3	1.7	
				4.7	6.8	2000	0.9	4.4	−2.8	2.7	3.1	3.8	−2.6	3.2	2.8	
C	8.0	2.0	1.0	1.65	6.4	400	1.2	3.7	−1.9	2.2	2.6	—	—	—	—	Berjot *et al.* (1970)

[a] k^2 is the lens strength, f_0 the focal length, $z(f_0)$ the distance from the lens center to the focal point, and c_s and c_c are, respectively, the spherical and chromatical aberration coefficients.

an average current density of 4×10^5 A/cm^2. In order to excite the ring, it is embedded in Co–Fe alloy (Figure 4.7). To achieve a higher field in the gap it was planned to replace the iron alloy by rare earth pole pieces (Hibino *et al.*, 1973).

Ring B consists of Nb–Sn sinter material with a shielding capability $\mu_0 H_{s1} = 6$ T/mm at the working temperature of 2 K. The excitation is carried out in an arrangement as sketched in Figure 4.5. The data for the axial field distribution measured with Hall probes were identical with those calculated by relation (4.1) within the limits of error.

The current-carrying capability of the Nb–Sn sinter material of ring C is relatively low, and so the data show no improvement compared to conventional lenses. The lens, however, should be used with a low k^2 so that the specimen is located above the bore hole. Lens C is well suited for this type of use.

The data for experiments carried out on a niobium ring lens (Boersch *et al.*, 1966*a*) are not given in Table 4.1 since the intrinsic critical current density of this material is poor.

4.1.4. *Iron Circuit Lens*

Since the half-width of the field distribution in the coils described in Section 4.1.2 is too large in many cases, iron shrouding is used to increase the peak field and to cut off the tails of the distribution. Thus in these lenses without pole pieces, H_0 arises from the combined effect of the coil field and the magnetization in the iron. Such iron-shrouded lenses have been extensively studied by Génotel *et al.* (1967) and Ishikawa *et al.* (1969). This type of lens is suitable for applications requiring beam voltages above 1 MV. Several designs and their fields along the z axis are sketched in Figure 4.8. Lenses D and E differ only in the size of the space within the iron circuit. Lens E may be used with a warm iron circuit; the solenoids are wound with Nb–Ti wire. In lens F the flux is generated by two coils, the main coil (driven in the persistent-current mode) and the shim coil (for focusing). The effective current density of the wire (Nb–Zr, diameter 0.2 mm) should not, to be safe, exceed 3×10^4 A/cm^2. Lens G is similar in design to D, E, and F, apart from the geometry of the iron

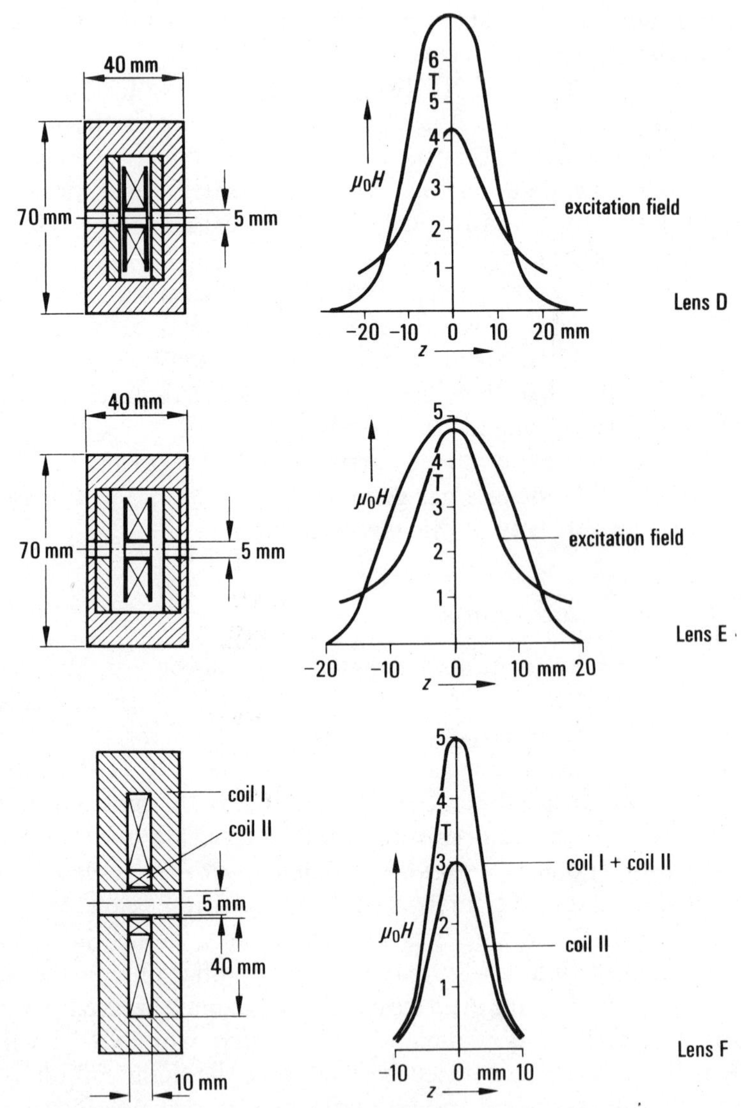

Figure 4.8. Iron-shrouded lenses and their field distributions.

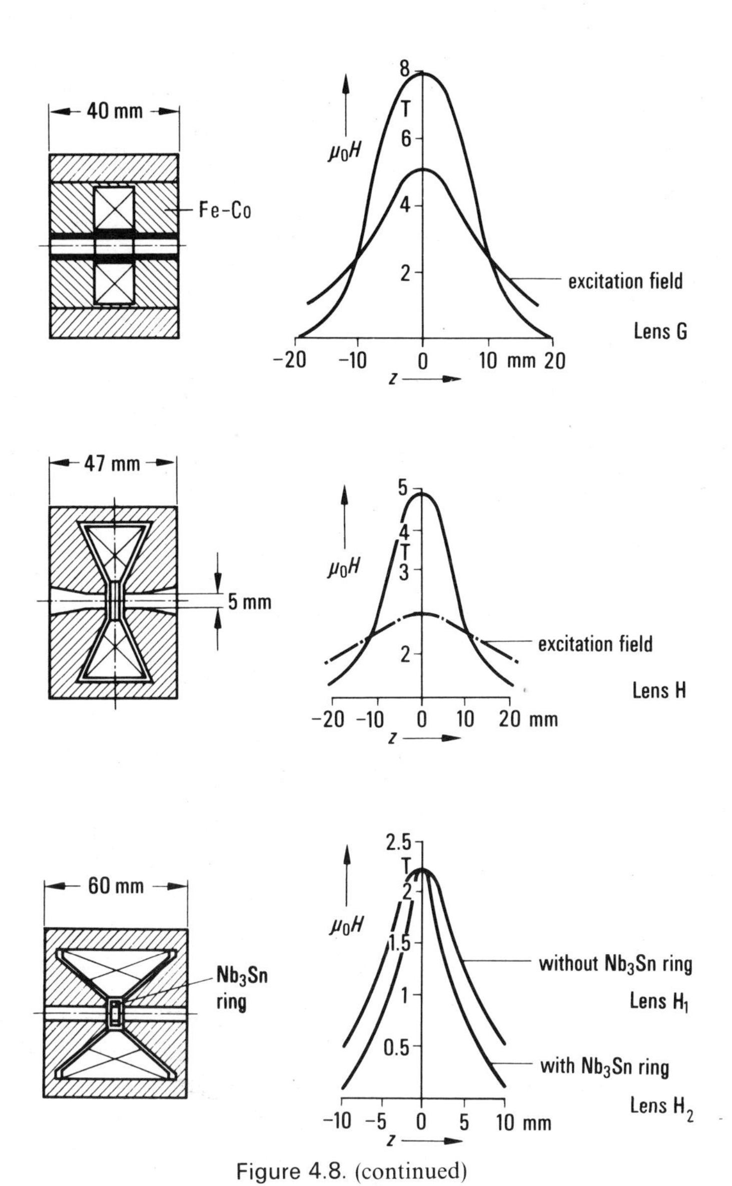

Figure 4.8. (continued)

circuit. Because of the high current density in the Nb–Ti wire used in lens G, the peak field is very high. In lens H, the iron circuit strongly determines the field distribution, so that we can consider this type as a transition to the pole-piece lenses. This is apparent if the excitation is diminished (lens H_1). In lens H_1 a ring made of Nb–Sn sinter material can be fastened in the gap to decrease the half-width (lens H_2). This combination can also be used in the ring lens mode. Cardinal elements and the aberration constants calculated for the field distributions of the lenses in Figure 4.8 are tabulated in Table 4.2.

The field distribution for a lens with unsaturated iron alloy pole pieces (peak induction lower than the saturation magnetization) is shown in Figure 4.9. The cardinal elements of the superconducting versions are quite similar to those of conventionally designed lenses. However, the lens size is scaled down as shown in Figure 4.10. Characteristic data for pole-piece devices

Table 4.2. Data for Iron-Shrouded Lenses

Lens	$\mu_0 H_0$, T	$2d$, mm	V, kV	k^2	f_0, mm	$z(f_0)$, mm	C_s, mm	C_c, mm	Reference
D	4.5	13.8	3000	1.8	5.4	−2.4	—	—	Génotel *et al.* (1967)
	5.6	14.1		3	4.6	−0.45	—	—	
	5.6	14.1	4000	1.8	5.7	−1	—	—	
	6.7	14.8		2.7	5.2	−0.9	—	—	
	5.6	14.1	5000	1.2	7	−4.3	—	—	
	6.7	14.8		1.8	6.1	−2.4	—	—	
E	5.0	23.2	4000	3.7	7	−0.2	—	—	
	5.0	23.2	5000	2.4	8.1	−2.3	—	—	
F	4.15	9.8	1000	4.9	2.3	—	1.5	2.0	Ishikawa *et al.* (1969)
	4.65	10	3000	1	5.4	—	4.8	4.1	
G	7.4	14.8	4000	3.4	4.7	0.13	3	3.5	Génotel *et al.* (1968)
	7.8	15.8	5000	3.1	5.4	0.39	3.75	3.9	
	5.1	10.4	3000	1.5	6.25	−2.2	—	—	
H	5.1	10.6	3000	1.5	6.25	—	—	—	Génotel *et al.* (1967)
	4.4	9.2	3000	0.9	6.55	—	—	—	
H_1	2.2	12	400	—	3.8	−2.05	1.4	2.4	Payen *et al.* (1971)
H_2	2.2	6.8	400	—	2.8	1.05	1.5	2.0	

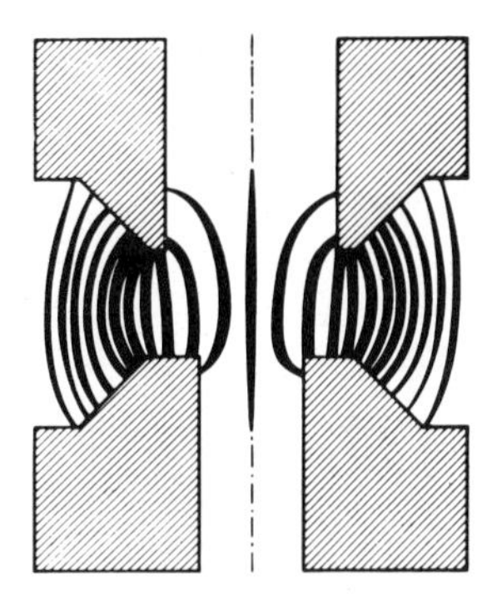

Figure 4.9. Field distribution between the pole pieces of a lens with unsaturated iron circuit. The width of the field lines is proportional to $|\mathbf{H}|$.

are summarized in Table 4.3. Since it is so easy to operate the conducting lenses with a high field, some authors also studied the pole-piece types in the saturated state.

The data for lens I are given for two different excitations. For the saturated device (I_1), the focal length is rather large. Lens K is similar to lens I, but has additional Nb_3Sn shielding near the bore to get rid of the tails of the field distribution. This favorably influences the focal length, but the chromatic aberration is somewhat higher than in lens I_2. Lenses L, M, and N are saturated. Lenses M and N will be discussed in Chapter 5 in connection with a superconducting 400-kV lens system, where they are equipped with warm iron circuits. Lens O was installed in an evaporator cryostat.

Higher peak fields can be obtained by using rare earth pole pieces. The field along the axis of such a lens (Figure 4.11), with its irregularities at the joints between the rare earth pole pieces and the iron yoke, is by no means bell-shaped. Figure 4.12 shows the voltage dependence of f_0, c_s, and c_c for lens O (Table 4.4) as calculated from the relation

$$H_z = \frac{H_0}{1 + (z/d)^n} + H_1 \tag{4.4}$$

Equation (4.4) is derived from a field model consisting of a modified Glaser field and triangular secondary peaks (peak field H_1). Optical parameters for another lens (P) of this type with optimum excitation are given in Table 4.4. The pole pieces of lens O are

Table 4.3. Data of Iron Pole-Piece Lenses

Lens	$\mu_0 H_0$, T	NI, At	$2d$, mm	$2s$, mm	r, mm	V, kV	k^2	f_0, mm	$z(f_0)$, mm	c_s, mm	c_c, mm	Reference
I_1	3.7	64000	—	4	2.1	3000	—	9.8	—	6.0	4.4	Trinquier and Balladore (1968)
I_2	1.8	—	5.6	4	2.1	300	1.5	3.4	—	1.9	1.4	
K	1.8	—	5.3	4	2.1	300	1.5	2.6	—	1.9	1.9	Trinquier *et al.* (1969)
L	—	30500	—	—	—	1000	3	7	—	3	5	
M	—	7700	—	—	—	300	—	6.5	—	4.95	5	
N	—	17000	—	3	3	400	—	3	0	1.4	2.2	Severin *et al.* (1971)
O	4	21800	7.0	1	1.8	1000	—	3	—	—	—	Verster and Kiewiet (1972)

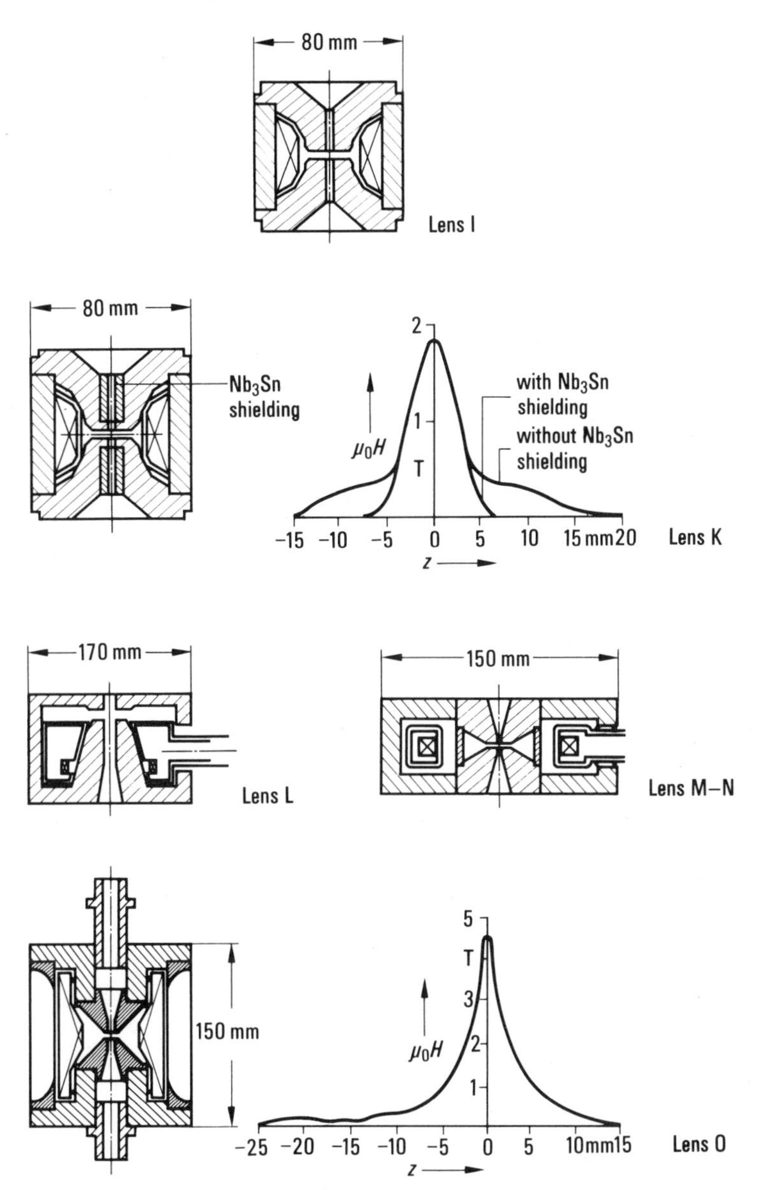

Figure 4.10. Iron pole-piece lenses and their axial field distributions.

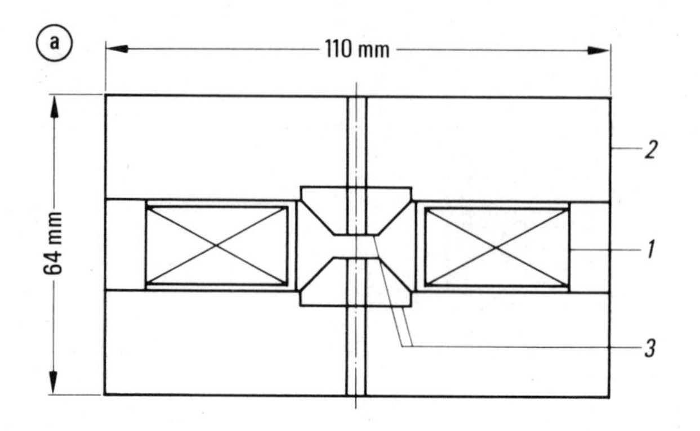

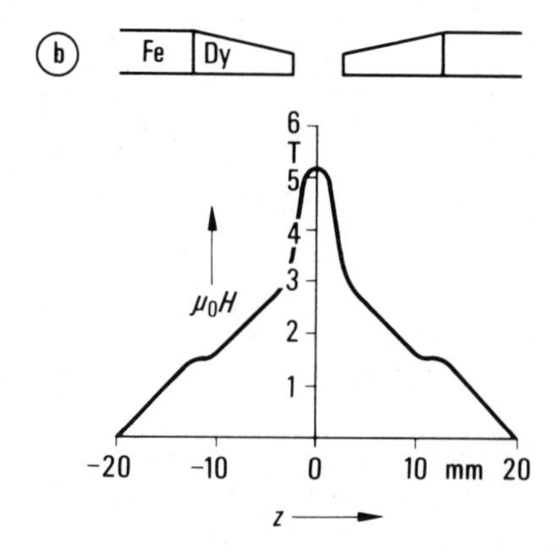

Figure 4.11. Lens with rare earth pole pieces. Reproduced by permission of Bonjour (1973). (a) Design: (1) coil; (2) iron circuit; (3) rare earth pole pieces. (b) Field distribution.

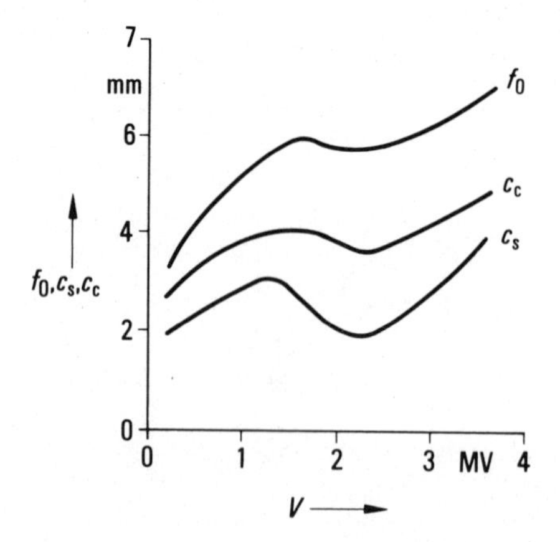

Figure 4.12. Focal distance f_0 and the chromatic and spherical aberration constants c_c and c_s of lens O (Table 4.4) as a function of the beam voltage. Reproduced by permission of Bonjour (1973).

Table 4.4. Data for Rare Earth Pole-Piece Lenses[a]

Lens	$\mu_0 H_0$, T	$2d$,[b] mm	s, mm	r, mm	V, MV	f_0, mm	c_s, mm	c_c, mm
O	5	5.66	3	2.2	2.8	9	2.7	5.5
P	6	—	2.5	1	3	7	2.4	4.4

[a] Bonjour (1973).
[b] $2d$ is the half-width of the central bell-shaped field.

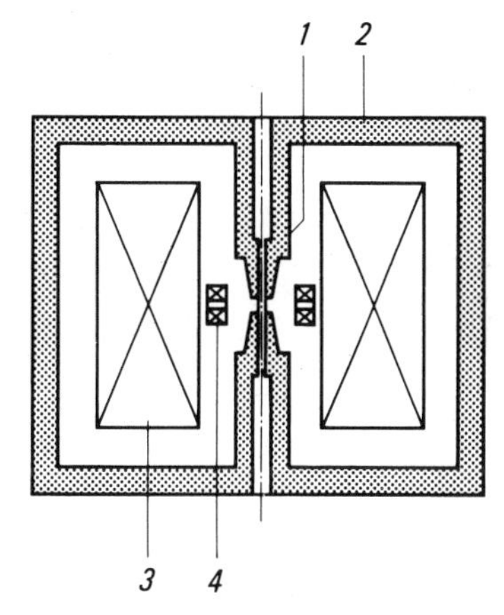

Figure 4.13. Design of the shielding lens: (1) shielding cylinders; (2) shielding container; (3) coils; (4) correction systems.

made of dysprosium, while those of lens P are of holmium. The design of lens O is asymmetric relative to the gap.

4.1.5. Shielding Lens

The shielding lens (Figure 4.13) is constructed without any ferromagnetic material (Dietrich *et al.*, 1969; Weyl *et al.*, 1972). It consists of a coil for field excitation, superconducting shielding cylinders made of Nb–Sn sinter material to concentrate the field in the gap, and a shield casing to prevent stray fields which might affect the electron beam. The field distribution (Figure 4.14) is in many ways inverse to that in a conventional lens. It was calculated by the Liebmann method assuming ideal shielding (Figure 4.14a) and using the Kim–Anderson relation to take into account the penetration of the field into the shielding cylinders in the region of the gap with increasing applied field (Figures 4.14b, c). The effective wall thickness decreases and the effective

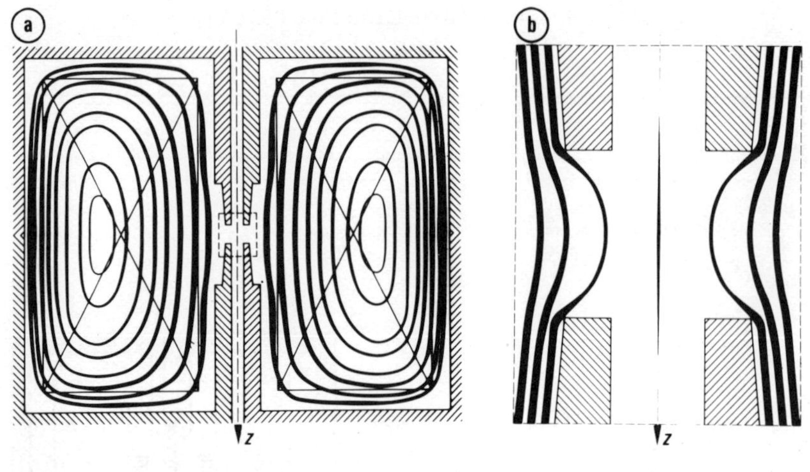

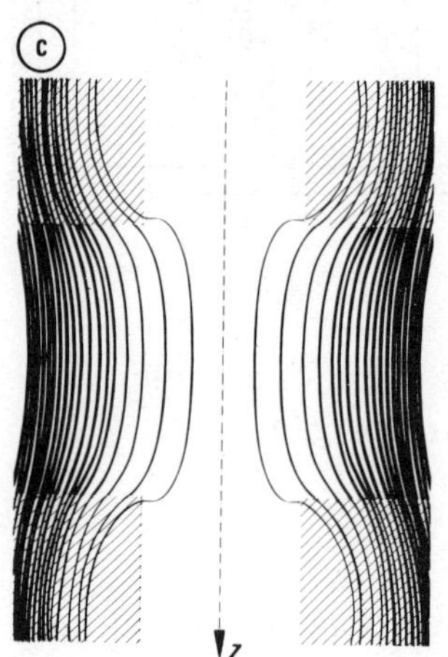

Figure 4.14. Field distribution in the shielding lens. (a) Calculated on the assumption of ideal shielding. (b) Enlarged gap region. (c) Calculated considering the field penetration. The width of the field lines is proportional to $|\mathbf{H}|$.

gap width enlarges as the critical state penetrates the superconductors. The axial field (Figure 4.15) can be approximately described by a Gaussian function

$$H_z = \exp[(z/d)^2 \ln 2]$$

for $-d \leq z \leq d$ and by a straight line in the region $z \geq d$ and $z \leq -d$.

Characteristic parameters for lenses operated with different field excitations are given in Table 4.5. The field penetration into

Table 4.5. Data for Shielding Lenses

Lens	$\mu_0 H_0$, T	$2d$, mm	V, kV	$2s$, mm	r, mm	Bell-shaped field approximation					Numerically calculated			
						k^2	f_0, mm	$z(f_0)$, mm	c_s, mm	c_c, mm	f_0, mm	$z(f_0)$, mm	c_s, mm	c_c, mm
Q	3.0	4	300	4	1.1	1.9	2.1	−0.7	0.8	1.3	1.6	−0.6	1.2	1.2
	3.0	4	400	4	1.1	1.4	2.2	−1.0	1.0	1.5	1.9	−1.1	1.4	1.4
P	6.7	5.5	1500	4	1	2.1	2.8	−0.6	1.0	1.8	2.2	−0.8	1.4	1.6
	6.7	5.5	2000	4	1	1.3	3.2	−1.5	1.5	2.8	2.7	−1.6	2.1	2.0
S	7.3	6.4	1500	4	1	3.3	3.2	−0.2	0.9	1.9	2.0	−2.0	1.3	1.5
	7.3	6.4	3000	4	1	1.0	4.2	−2.5	2.4	2.8	3.4	−2.5	3.2	2.6

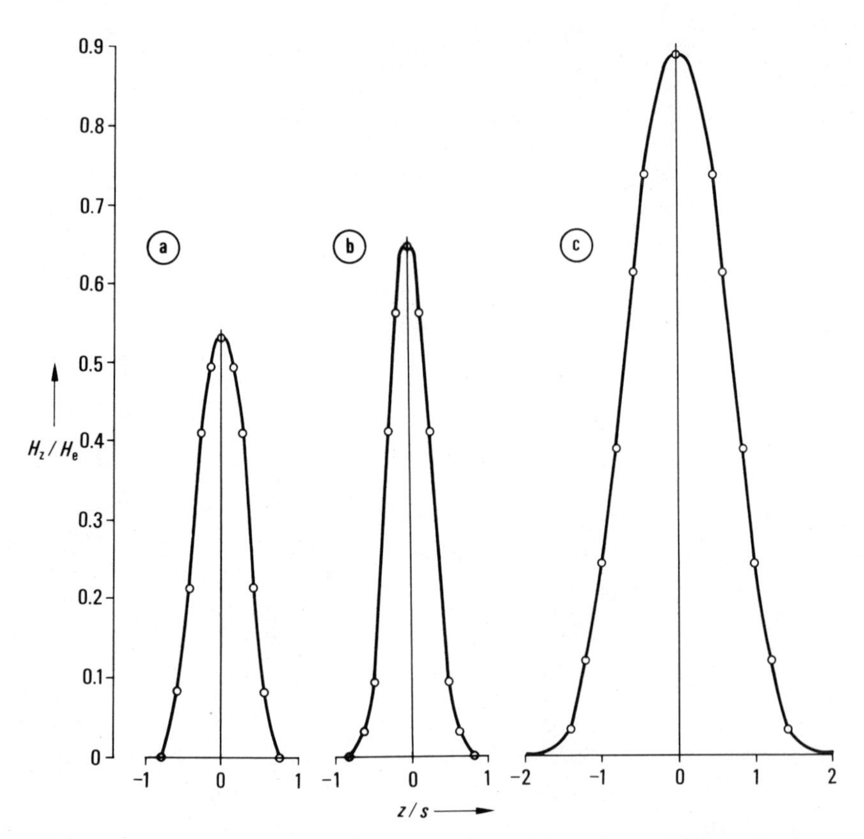

Figure 4.15. Axial field distribution in the shielding lens due to field penetration. (a) Ideal shielding (calculated curve). (b) Nearly ideal shielding (experimental curve, $\mu_0 H_e = 1.8$ T). (c) Enlarged effective gap width due to field penetration (experimental curve, $\mu_0 H_e = 8.2$ T). $2s$ is the gap width and H_e the external field.

the shielding cylinders of lens Q is insignificant, while in lenses R and S the effective gap width has been increased since the critical state extends farther into the bore than is shown in Figure 4.14.

4.1.6. Advantages and Drawbacks of the Different Lens Types

In our evaluation of various types of lenses, simple solenoids are not included since there are no plans to use them in a microscope. The experiments of Fernández-Móran with a coil lens are of only historical interest today. It might yet prove possible to install a specimen stage with a lens of the Merli type. Applications of coils in electron optics not connected with microscopy are discussed in Appendix A.

Iron circuit lenses have the enormous advantage that the experience gained in classical electron microscopy can be used, e.g., the specimen stages correspond more or less to the well-known prototypes. The lenses can be focused with the objective current either directly or with an inductive method if the coils are driven in the persistent-current mode. The lens dimensions are rather small. On the other hand, if really high fields ($\mu_0 H \geq 3$ T) are used, either the lens has to be constructed with rare earth pole pieces (not much is known yet about the anisotropy of this material) or one has to make use of the shrouded-lens type, where the field is produced mainly by the coil. In both cases there might be large deviations from rotational symmetry of the field. Besides this, the d values are relatively large, so that if the lens strength does not exceed $k^2 \approx 2$, the lenses are only adaptable to very high beam voltages.

At first sight the ring lens is a simple device, and if a superconductor with a sufficiently high current density is available, a small half-width and a correspondingly short focal length at high fields can be obtained. On the other hand, the lens is like a permanent magnet, and if a new adjustment is necessary, the lens has to be warmed up. Heating the ring by the electron beam may cause flux jumping, which means deterioration of the field. Focusing is only possible by varying the voltage. It may be possible to perform small changes of the peak field with an additional coil within the bore hole of the ring, although this has not been experimentally verified. Such a ring might produce strong deviations from rotational symmetry, a possibility that can probably be minimized by constructing it with a stack of disks (Table 4.1, lens A). Such a design permits only a top-entry specimen stage if $k^2 > 1$, a disadvantage if the specimen must be tilted.

Because of the steep slopes of the axial field distribution, the shielding lens has an even smaller focal length than the ring lens if the same superconducting material is used. The spherical and chromatic aberration constants are also small. Some of the problems resemble those of the ring lens. Since lowering the excitation field produces a strong hysteresis effect, which is a function of the peak field, focusing is better done by adjusting the voltage. The lens is sensitive to flux jumps. The deviation of the field distribution from the desired rotational symmetry may be dangerous, but it can be corrected with deflection and stigmator coils in the gap, as pointed out in the next chapter. The construction is facilitated if a top-entry specimen stage is replaced by one with side entry since the latter has a number of advantages.

4.2. Correction Systems for Superconducting Objective Lenses

Because of their reduced size, superconducting deflector and stigmator coils may be placed in the gap of the lens. They are effective in saturated iron circuit lenses and especially in shielding lenses. In unsaturated iron lenses, the flux of the correction coils is drawn into the pole pieces. In the ring lens, the space in the bore is too restricted to permit the use of correction coils.

The operation of such compensation systems has been studied extensively for misaligned shielding lenses to determine the extent to which a mechanical adjustment can be replaced by an electromagnetic one (Knapek, 1975). The situation in the gap of the shielding lens was simulated with an electrolytic tank, and the course of the potential, and hence the field lines, was determined. In the enlarged model (Figure 4.16), two insulating Araldite cylinders (A) represent the shielding cylinders; two parallel brass disks (B) with their normals aligned with the z axis simulate the field coil; and near the gap, two brass disks (C) oriented with their normals perpendicular to the optical axis substitute for the correction coils. The determination of the equipotential lines was controlled by an analog computer. The curves were plotted on a coarse mesh net. To derive the field lines, the potential values were determined at each point and the field direction and magnitude were calculated by differentiation.

In Figure 4.17, the course of the field lines in the gap of a

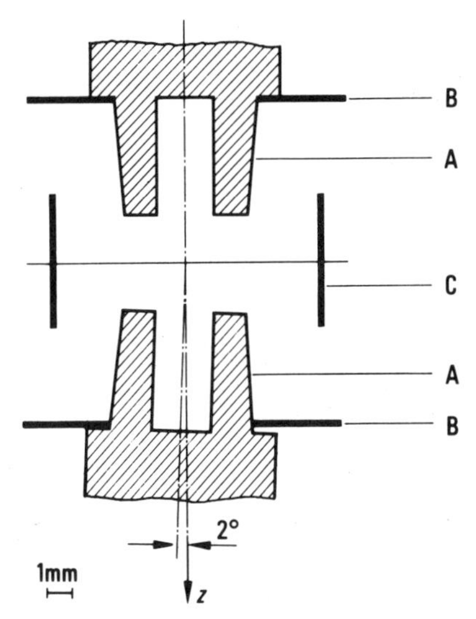

Figure 4.16. Model lens in an electrolytic tank: (A) Araldite cylinders (upper cylinder slightly tilted); (B) brass plates for simulation of main coils; (C) brass plates for simulation of correction systems.

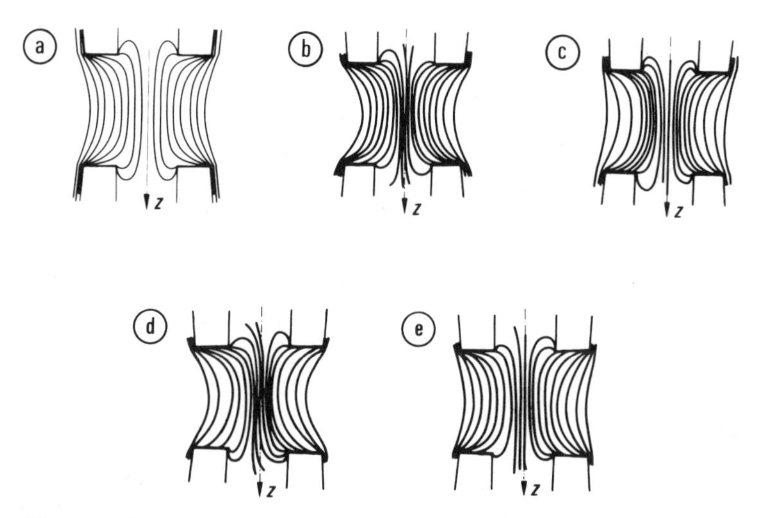

Figure 4.17. Field lines in the gap of a shielding lens. (a) Well-aligned lens. (b) Lens with displaced cylinders (displacement 112 μm corresponding to scale in Figure 4.16). (c) Lens with displaced cylinders and correction field. (d) Lens with tilted cylinders (tilt angle 2°). (e) Lens with tilted cylinders and correction field.

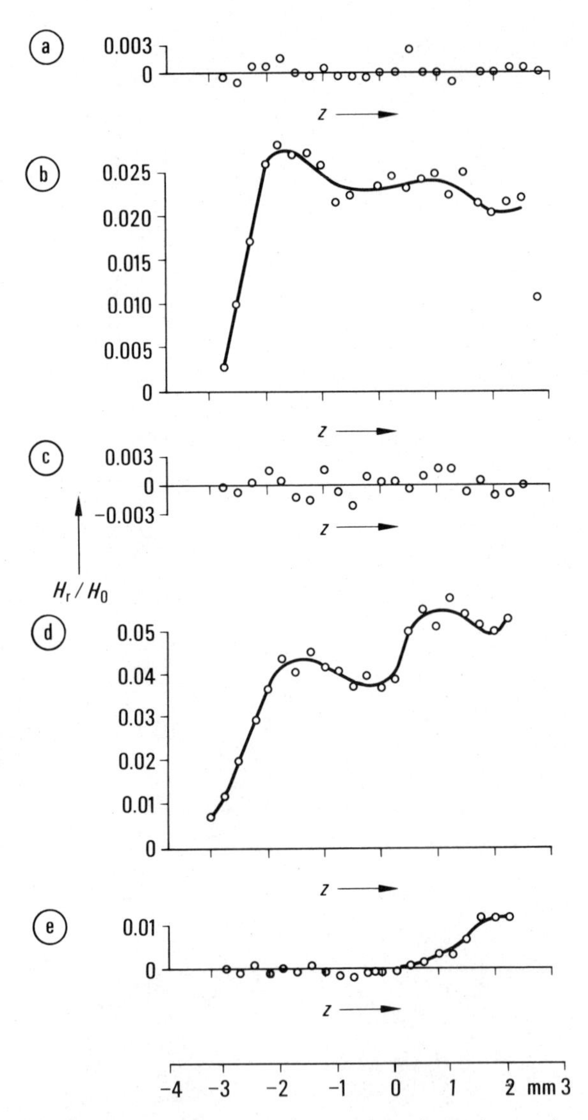

Figure 4.18. Radial components H_r along the z axis for the field distribution of Figure 4.17. (a) Well-aligned lens (deviations are caused by errors in measurement). (b) Lens with displaced cylinders (displacement 112 μm). (c) Lens with displaced cylinders and correction field. (d) Lens with tilted cylinders (tilt angle 2°). (e) Lens with tilted cylinders and correction field.

well-aligned shielding lens (Figure 4.17a) is compared to that in lenses with displaced shielding cylinders (Figure 4.17b) and tilted shielding cylinders (Figure 4.17d). The effect of compensation coils is shown in Figures 4.17c, e. The transverse field components for these five cases without and with correction coils are shown in Figure 4.18. Calculations similar to those of Sturrock (1951), Archard (1953), and Riecke (1973) make it obvious that for the scale assumed here, a cylinder displacement of 112 μm can be compensated for, so that the resolution is not affected. A 2° tilt can be corrected up to a small residual, which causes a misalignment coma. This disturbance should be negligible.

Thus it has been shown that large deviations from rotational symmetry in the original field distribution are permissible if correction systems are used. More details on the design of dipole and quadrupole magnets used for correction systems are given in Chapter 5.

4.3. Testing of Objective Lenses

4.3.1. Testing Objective Lenses with Simulation Devices

To get an approximate idea of the image aberrations of an objective lens, a method proposed by Leisegang (1953) can be used. This method does not need an elaborate cryostat adapted to a microscope (Dietrich *et al.*, 1970). In addition, a high beam voltage up to 10^3 kV can be simulated with 100 kV.

An electron optical bank suitable for this purpose is shown in Figure 4.19 (Knapek, 1976). The cathode spot is imaged by condensers I and II. The crossover situated near the valve replaces the specimen. The objective lens, immersed in a metal cryostat, is operated in a telecentric mode with $k^2 \approx 3$ or $k^2 \approx 8$ (two crossovers) in the Glaser limit depending on the voltage to be simulated. If there is a transverse component of the field along the z axis, the beam is deflected and the caustic figures on the screen are displaced (the caustic is the envelope of the focused ray). The displacement values permit calculation of the strength of the transverse field in the gap. Typical results for the shielding

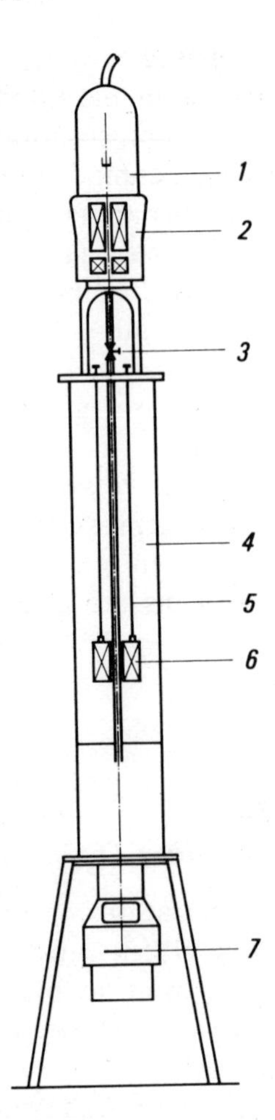

Figure 4.19. Electron optical bank for lens testing: (1) electron source; (2) condensers I and II; (3) valve; (4) cryostat; (5) rods for alignment; (6) objective lens; (7) screen.

lens are given in Figure 4.20, where the ratio of the deflection angle γ to the tolerance angle γ_0 for a resolution of the order of 0.1 nm is plotted as a function of the beam voltage. If there is no change in the lens strength, the peak field must be raised with increasing beam voltage, and this causes an enhanced penetration of the field into the shielding cylinders. As the penetration is not quite uniform, γ increases.

The same is true for the astigmatism. If there is an elliptic deviation from the circular shape of the shielding currents, an axial astigmatism is produced with a strength which can be

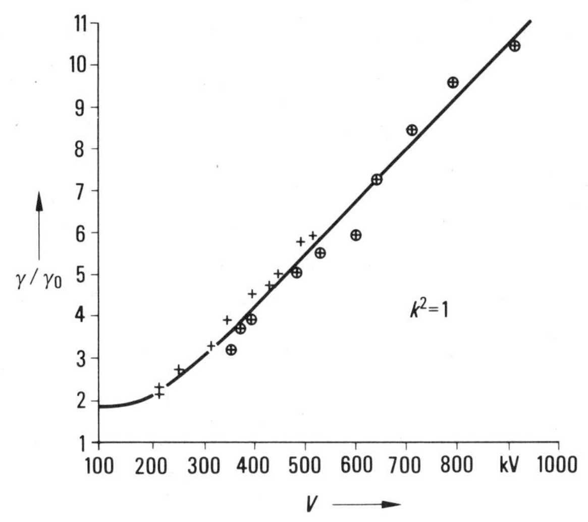

Figure 4.20. Deflection angle ($\tan\gamma = H_r/H_0$) as a function of the beam voltage. The measurements were carried out in the telescopic mode of the lens with $k^2 = 3$ and $k^2 = 8$ in the Glaser approximation, and the results calculated for $k^2 = 1$; γ_0 is defined as the deflection angle permitting a resolution of ≈ 0.1 nm; $\gamma_0 = 5 \times 10^{-4}$ rad: (+) Glaser approximation, $k^2 = 3$; ($\oplus$) Glaser approximation, $k^2 = 8$.

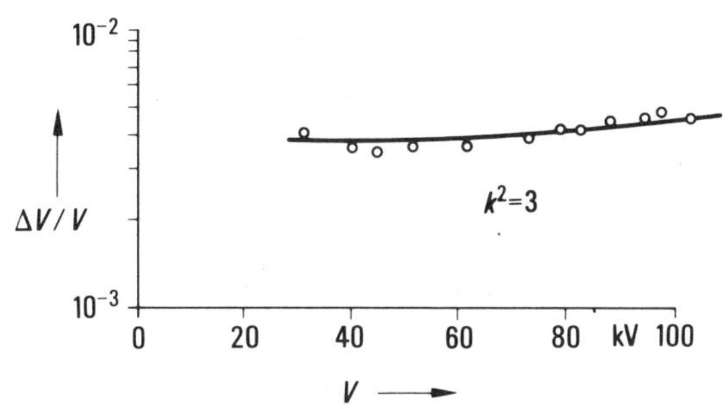

Figure 4.21. Astigmatic difference ΔV normalized to V as a function of V for the telescopic lens mode. The beam voltage adequate for smaller lens strengths can be calculated from equation (2.1). The tolerance for the astigmatic difference for a resolution of 0.1 nm is estimated as $\Delta V/V \leq 10^{-4}$. Glaser approximation, $k^2 = 3$.

obtained from the distance between the upper and lower caustic points. It is measured by the voltage difference ΔV for both adjustments, normalized to the beam voltage V. The basic astigmatism as a function of the voltage ($k^2 \approx 3$) is shown in Figure 4.21. The results given in Figure 4.20 and 4.21 exceed by only an order of magnitude the tolerance limits of astigmatism and deflection for a resolution of 0.1 nm even in the high-voltage region. The experiments presented in Section 4.2 prove that in the shielding lens larger deviations can also be easily compensated for with correction systems.

For simulating high voltages, Bonjour (1973) used, instead of electrons, lithium ions with identical $Ze/(mV^*)$ for imaging and thus could imitate beam voltages of 3 MV and 5 MV by 1 kV and 2.3 kV, respectively. The experiments were carried out in a device (Figure 4.22) with a heated silicate ball used as an ion source, an electrostatic condenser for beam focusing, and an objective lens (Table 4.4, lens P), but no magnifying lenses. The specimen consisted of a grid. When the ion current was small the

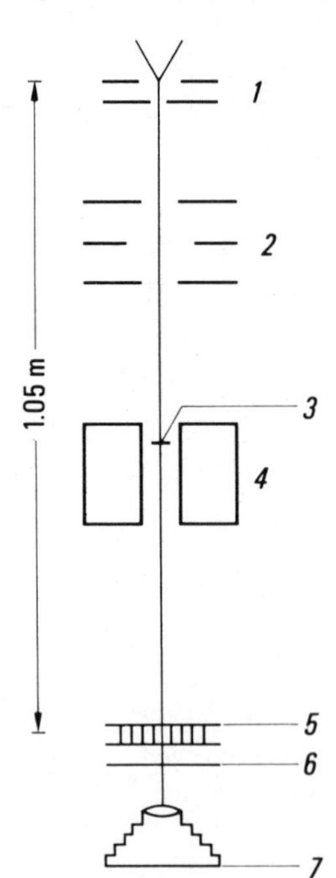

Figure 4.22. Device for lens-testing simulating high beam voltages. Reproduced by permission of Bonjour (1973): (1) Ion source; (2) electrostatic condenser; (3) specimen; (4) objective lens; (5) channel multiplier; (6) screen; (7) camera.

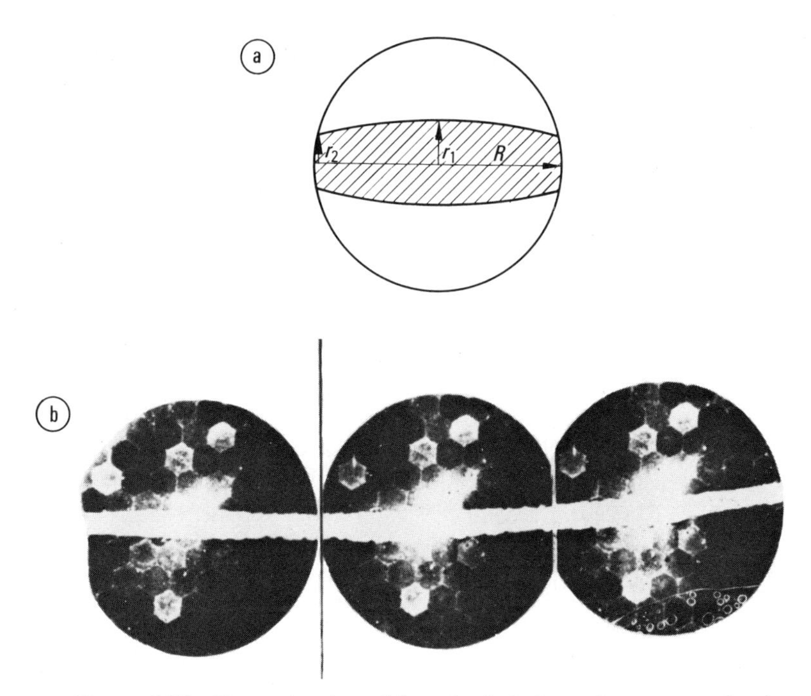

Figure 4.23. Determination of the spherical aberration constant by the method of Dosse. Reproduced by permission of Bonjour (1973). (a) Principle of the method [see equation (7.1)]. (b) Test micrograph.

ionic image was processed by a channel multiplier with a magnification that varied between 1000 and 10,000. From the image of the grid and the distorted shadow of a very slender wire (Figure 4.23), the spherical aberration coefficient could be obtained using the method of Dosse (1941):

$$c_s = rD^3 \frac{(1/r_2) - (1/r_1)}{D^2 - r_1^2} \tag{4.5}$$

where r is the radius of the wire, r_1 and r_2 the radii of the wire image in the center and near the edge, respectively, and D the distance of the object from the channel multiplier. The agreement with the calculated values (Table 4.4) was reasonable, with deviations of 10–30%.

4.3.2. Testing in the Microscope

At present the objective lenses have been tested in the electron microscope with beam voltages only up to 350 kV. A survey is given in Table 4.6. Some micrographs taken with the lenses listed in Table 4.6 are reproduced in Figures 4.24 and 4.25.

An excellent current stability was maintained as all the lenses were operated in the persistent-current mode. Focusing was carried out either by changing the voltage or by using superconducting transformers for small current changes. However, the experimental resolution values do not yet approach the theoretical limit. This is partly due to mechanical vibration and drift, and partly to unstable high-voltage sources. The voltage stability in most cases amounted to $\Delta V/V > 5 \times 10^{-6}$, with the exception of the device of Worsham *et al.* (1974). Here a stability of $\Delta V/V = 3 \times 10^{-7}$ is reported. But since even with a field emission gun the energy spread of the voltage is about 0.3 eV, a smearing of the electron energy of 2 ppm cannot be avoided for the usual beam voltages, i.e., below 150 kV. Information on specimen drift is not available. According to Fernández-Móran (private communication), the drift could be reduced by cooling in superfluid helium.

Table 4.6. Resolution Tests of Superconducting Objective Lenses

Lens type	V, kV	k^2	f_0, mm	Point resolution, nm	Lattice resolution, nm	Reference
Iron core	100	—	—	1.0	—	Fernández-Móran, (1970)
Iron core warm (lens)	100	—	—	0.4	—	Bonhomme et al. (1973)
Iron core warm (lens L)	300	—	6.5	1.0–1.5	—	Trinquier *et al.* (1972)
Iron core (lens F)	100	—	—	—	1.2	Ishikawa *et al.* (1969)
Iron core	150	6	2	0.4	—	Worsham *et al.* (1974)
Ring	40	—	4.3	10	—	Boersch *et al.* (1966*a*)
Ring (lens A)	75	2	1.7	1.0	—	Hibino *et al.* (1973)

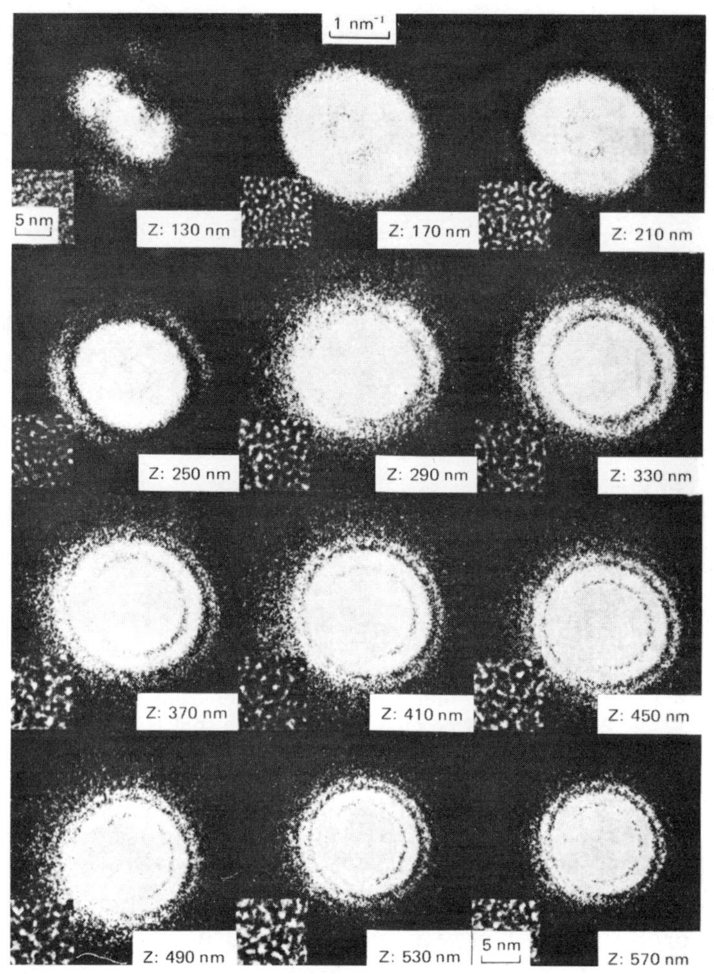

Figure 4.24. Slightly defocused micrographs of carbon foils (inserts) and their light optical diffractograms. Reproduced by permission of Bonhomme *et al.* (1973). According to the phase contrast transfer theory, there should be extinction in the diffractograms at positions depending on the defocusing distance Z (distance from the focal point) and the wavelength. The spatial frequencies are transferred up to a certain limit determined by the aberration constants, the energy spread, the aperture angle, etc. If this threshold value is exceeded, the contrast decreases and the rings disappear. According to the scale at the top of the figure, in the diffraction pattern for Z = 330 nm there are rings resolved up to a distance of 2.5 nm^{-1} in the reciprocal space, corresponding to a resolution of 0.40 nm. A normal-conducting lens with similar cardinal elements in the same microscope exhibited a resolution of 0.48 nm.

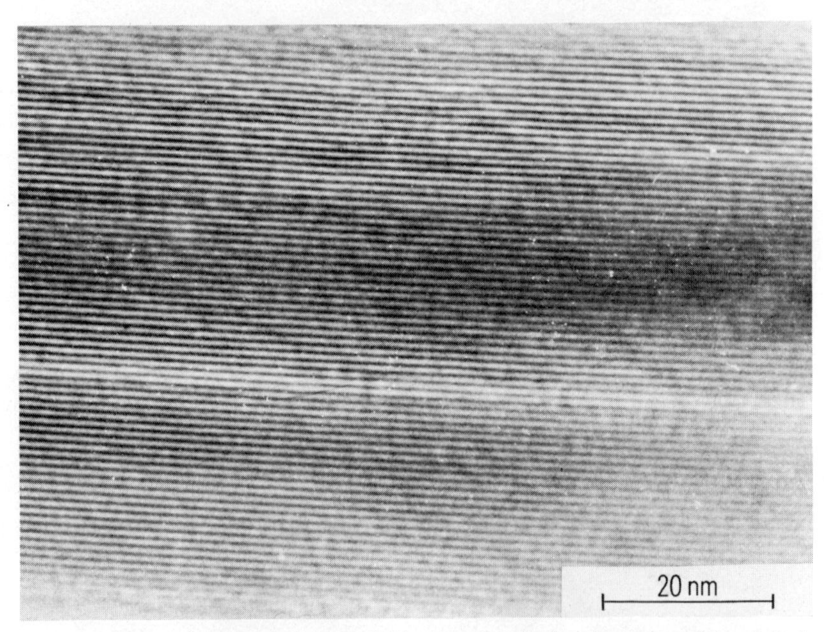

Figure 4.25. Lattice image (1.2-nm lattice constant) of a clay mineral (sepiolite). Reproduced by permission of Ishikawa *et al.* (1969).

Much of the lens data in Tables 4.1–4.5 are calculated for the one-field-condenser mode ($k^2 \approx 3$) for various reasons, in particular to present small error constants. In the experiments, a lens strength $k^2 \approx 1.5$ was preferred, mainly to avoid the adjustment problems which arise in one-field-condenser objectives.

5 Systems with Superconducting Lenses

5.1. Tested Systems

Superconducting lenses have not been much used yet as condensers, intermediate lenses, and projectors in conjunction with superconducting objective lenses. Projectors are used, as in conventional electron microscopy, to magnify the intermediate image produced by the objective lens. Since the magnification should be a maximum for a given column length, the lens strength with the smallest asymptotic focal length should be used. This leads to $k^2 \approx 1$ in the bell-shaped field approximation. For lenses with a field distribution with steep slopes, the asymptotic focal length increases less as the excitation increases. An example is the shielding lens R (Table 4.5 for the case of a 1500-kV beam voltage) for which $f_1 \approx f_0$ was found by numerical calculation, although the Glaser value of k^2 in the bell-shaped field approximation corresponds to $k^2 = 2.1$. But since the magnification should be controlled with the intermediate and projector lenses, the strong hysteresis in the shielding lens is very inconvenient and thus only iron-circuit lenses have been used up to now for further magnification.

A microscope with a column consisting of five superconducting lenses with warm iron circuits and Nb–Ti single-core wire coils is being operated by the Paris group (Severin *et al.*, 1971; Laberrigue *et al.*, 1974*a,b*). It is designed for beam voltages

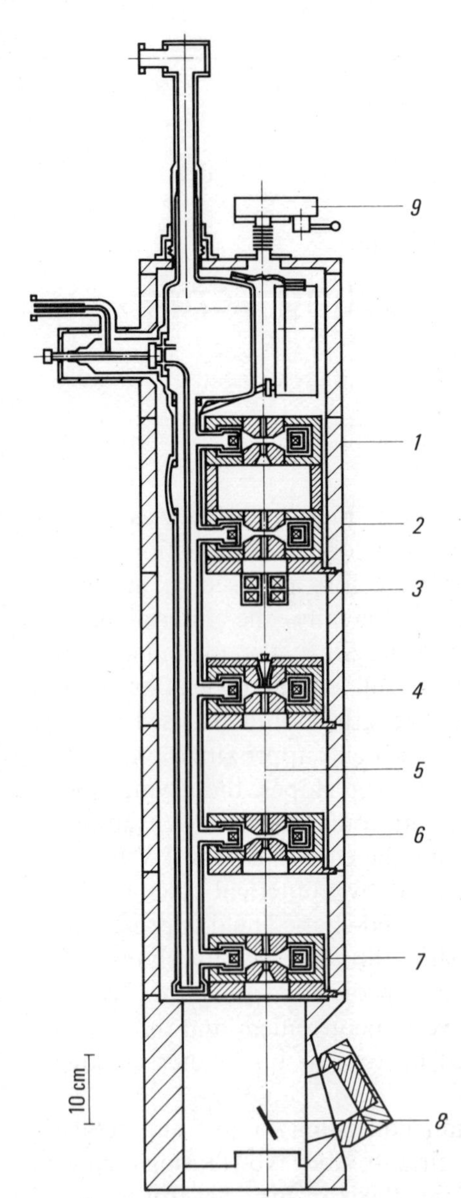

Figure 5.1. 400-kV microscope with superconducting lenses. Reproduced by permission of Laberrigue *et al.* (1974*a*). (1) Condenser I; (2) condenser II; (3) two-stage deflector; (4) objective lens; (5) room for further lens; (6) intermediate lens; (7) projector; (8) observation chamber; (9) lock between column and accelerator.

Table 5.1. Lens Dimensions in mm for the System Shown in Figure 5.1

Magnitude	Condenser I	Condenser II	Intermediate lens	Projector
s	3	8	8	3
r	1.5	4	6	2.3

up to 1000 kV and is at present equipped with a 400-kV generator (electrostatic device, fabricated by Tunzini Sames). Figure 5.1 shows the arrangement. The cryostat is described in Chapter 4. The condenser lenses, the intermediate lens, and the projector are lens types N or M (Table 4.3). The coils are symmetric relative to the midplane. The dimensions of the lens gaps s and bore radii r are given in Table 5.1. One power supply feeds all the lenses since a series circuit is used with persistent-current switches, as shown in Figure 3.7. A resistance $< -10^{-11}\ \Omega$ for the soldered contact in the superconducting mode is reported. Since there are strong attractive forces between the warm iron circuits and the coils at helium temperatures, especially for the objective lens constructed asymmetrically relative to the midplane, glass fiber mounts are provided in the cryostat for compensation. The details of the objective lens are shown in Figure 5.2. The specimen is inserted through a top entry with a goniometer stage. A side-entry specimen state is also planned.

The upper limit to the magnification achieved at 400 kV is 350,000. Because of the decreasing intensity, 200,000 is the maxi-

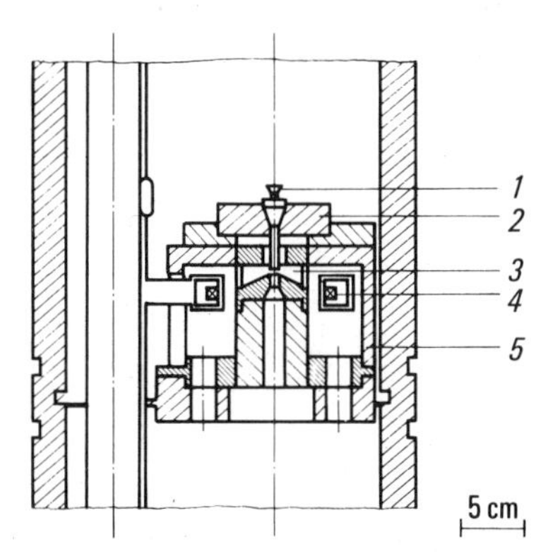

Figure 5.2. Objective lens of 400-kV microscope. Reproduced by permission of Laberrigue *et al.* (1974*a*). (1) Specimen carrier; (2) goniometer; (3) median plane of lens; (4) superconducting windings; (5) magnet circuit.

mum magnification that allows adjustment without an image intensifier. The instrument can be used 35 h a week with a helium consumption of 25 liters (the contents of a normal helium storage vessel). A number of published high-quality micrographs taken with this instrument indicate a resolution of 1 nm (Figure 5.3).

A small two-lens system was tested with beam voltages up to 400 kV (Dietrich *et al.* 1972*a*, 1973, 1974*b*). The system, as shown in Figure 5.4, is mounted between the double condenser and the second intermediate lens of an otherwise conventional microscope. The cryostat is described in Section 3.1.1. The system consists of a two-stage deflector; a shielding lens used as an objective lens, including stigmator and deflector in the gap and an improvised top-entry specimen stage; a further one-stage deflector; and a first intermediate lens. The deflection systems permit correction adjustments without the help of mechanical devices. The two-stage deflector allows the beam to be centered on the objective lens axis. In both planes there are eight coils of 50-μm Nb–Ti single-core wire. The electric circuit permits the beam to be shifted or tilted independently. A Nb–Ti shielding protects the beam from stray fields.

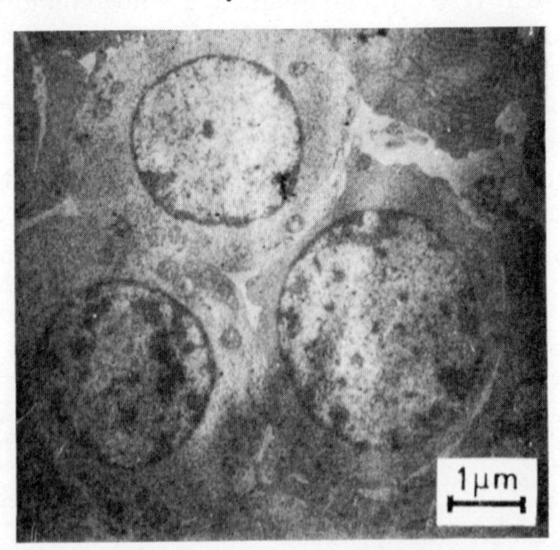

Figure 5.3. Micrograph taken with 400-kV microscope: section of fetal rat liver, $V = 300$ kV. Reproduced by permission of Laberrigue *et al.* (1974*b*).

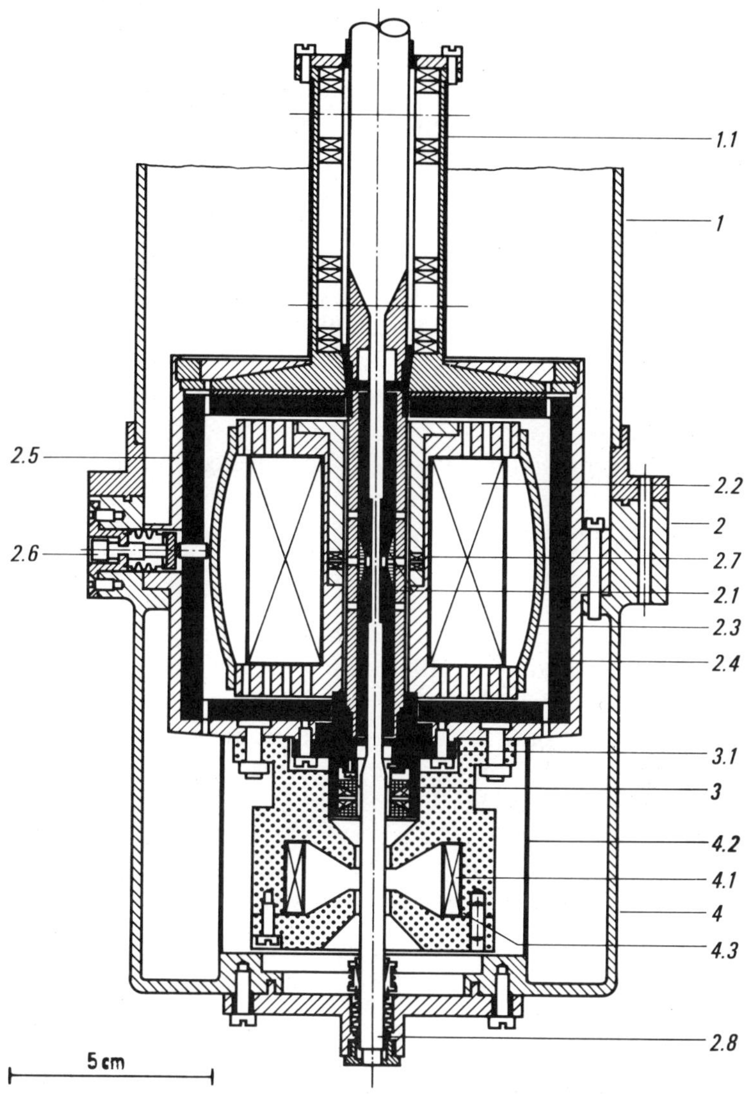

Figure 5.4. Superconducting imaging system: (1) two-stage deflector; (1.1) shielding; (2) objective lens; (2.1) superconducting shielding cylinders; (2.2) solenoid; (2.3) support for coils; (2.4) shielding container; (2.5) casing; (2.6) adjustment screw; (2.7) stigmator; (2.8) vacuum tube; (3) one-stage deflector; (3.1) Nb–Ti shielding; (4) intermediate lens; (4.1) solenoid; (4.2) superconducting shielding; (4.3) iron circuit; (■) superconducting material; (▩) ferromagnetic material.

The relevant data for the shielding lens are given in Table 4.5 (lens Q). The conically formed shielding elements, with a misalignment which does not exceed 1 μm, are arranged near the axially aligned vacuum tube. The superconducting coil is fastened to a support which can be tilted by adjustment screws. The shielding casing is mounted in a brass container for safety reasons. A stigmator made of eight Nb–Ti coils oriented with their axes perpendicular to the z axis allows correction of an astigmatic focal length difference $(\Delta f_0)/f_0 \leq 10^{-2}$. The same coil arrangement can be used for deflection with an appropriate electronic control. The improvised specimen holder is shown in Figure 5.5. Only a screwlike motion of the specimen is possible with this construction.

The intermediate lens is tightly connected with the objective lens to obtain the desired mechanical stability. Since the tolerances of the mechanical alignment of the two lenses cannot be kept below 10 μm, the fine adjustment has to be done with the one-stage deflection device. The intermediate lens is equipped with a cold iron circuit with the pole-piece design similar to that of Durandeau and Fert (1957) but adapted to a 400-kV beam voltage. The superconducting windings are made of Nb–Ti multicore wire. Some data for the intermediate lens are: diameter of the iron circuit 60 mm, bore diameter 9 mm, excitation 8000 At, asymptotic focal length $f_1 \approx 4$ mm.

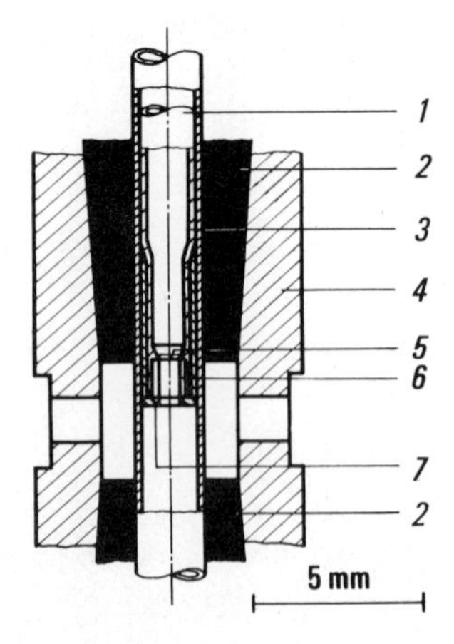

Figure 5.5. Provisional specimen stage for the system shown in Figure 5.4: (1) vacuum tube; (2) shielding cylinders; (3) tube; (4) mounting holder; (5) specimen net; (6) ring; (7) cap. (■) Superconducting material.

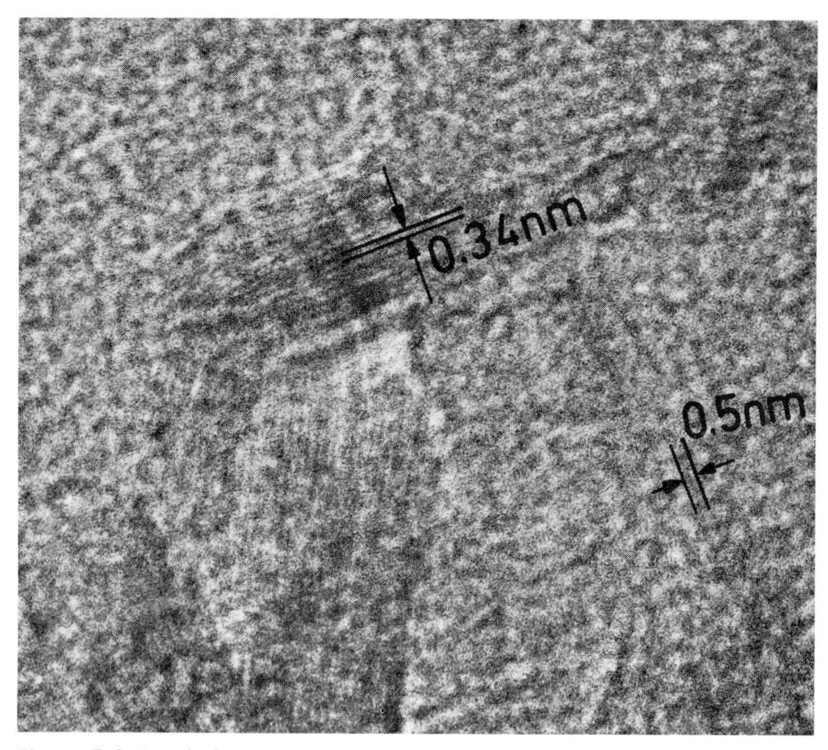

Figure 5.6. Resolution test taken with the system shown in Figure 5.4. Graphite crystals on carbon foil, $V = 350$ kV. The distance between the graphite lamellas is 0.34 nm, and the point resolution in the amorphous carbon is < 0.5 nm (as tested with the method used in Figure 4.24).

A point resolution below 0.5 nm was achieved with this system (Figure 5.6). This result is confirmed by an evaluation with a light optical diffraction method (Thon, 1966; Hoppe *et al.*, 1970), which demonstrates that space frequencies > 2 nm^{-1} are transferred. The images of graphite crystals indicate a lattice resolution of 0.34 nm. The resolution is probably mainly limited by the instabilities of the high-voltage source and mechanical instability. The drift could be reduced to 0.3 nm/min during the warm-up period of the helium bath after the temperature has been lowered to ≈ 2.5 K. The micrograph shown in Figure 5.6 was taken with an exposure time of 6 sec.

After achieving these results, a system with four superconducting lenses has been constructed (Figure 5.7) for installation in the cryostat shown in Figure 3.2 (Dietrich *et al.*, 1974*a*). An objective lens of the shielding type is provided with a side-entry specimen stage. The center of the lens is a stainless steel disk; double coils for the field excitation, the shielding cylinders, and the specimen and aperture slide rods are attached to the disk. To compensate for beam deflection, the lower-field coil section can be shifted transversely to the optical axis. The fine adjustment can be performed electromagnetically by the deflection coils.

The stigmator consists of eight coils. There is only a small space available for the correction devices, and they must be mounted as close as possible to the optical axis since they must be highly efficient. For this reason calculations were made for several coil designs and compared. In the upper part of Figure 5.8 the arrangements are drawn in perspective, with the corresponding fields in the x-y plane outlined below them. The simplest arrangement with the lowest efficiency consists of elliptical cylinder coils (type 1). The axes of the coil ellipses coincide with the x and y axes, respectively. Somewhat more efficient are four radially

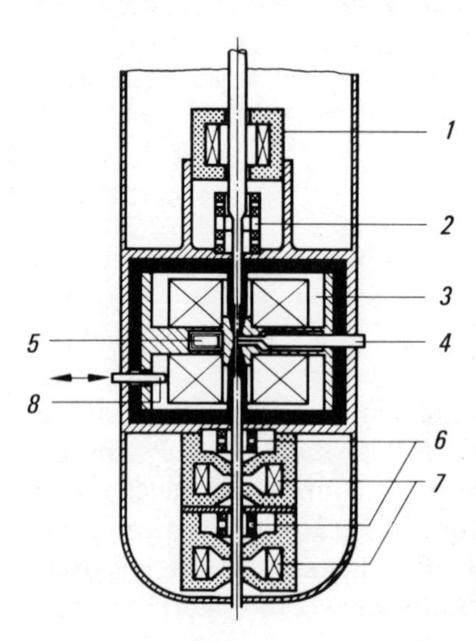

Figure 5.7. Superconducting lens system with side-entry specimen stage: (1) condenser; (2) two-stage deflector; (3) shielding; (4) specimen rod; (5) correction system; (6) deflectors; (7) intermediate lenses; (8) adjustment screw; (■) superconducting material; (▩) ferromagnetic material.

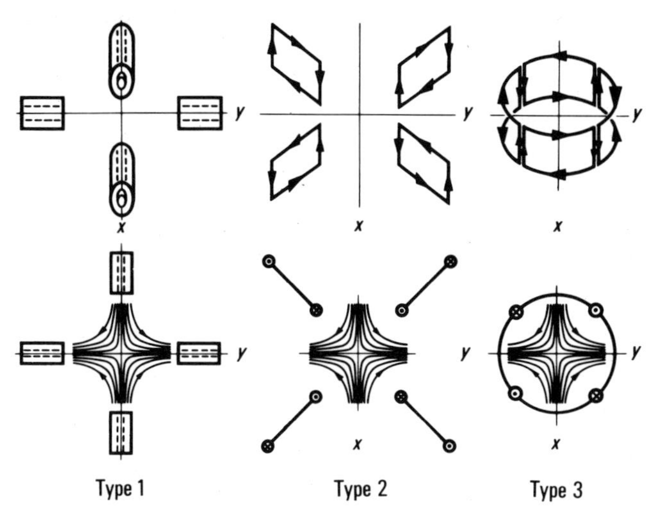

Figure 5.8. Stigmator designs with corresponding field distributions. Type 1, elliptical cylinder coils; type 2, meridional stigmator; type 3, azimuthal stigmator.

arranged rectangular coils with windings in the meridional planes (type 2). The front vertical sides of the conductors generate the field. The horizontal and rear vertical conductor parts reduce this field by 50%. In the third design, rectangular coils are slightly bent and arranged in the shape of a cylinder around the optical axis (type 3, aximuthal stigmator). In this case all the conductor parts contribute to the quadrupole and for this reason the azimuthal stigmator has the highest efficiency if the evaluation is based on the ampere-turn number.

Since the meridional system (type 2) needs less space and is easier to build, it was preferred. Figure 5.9 presents a cross section. The eight stigmator coils cannot be distributed equiangularly around the gap because of the holes for the specimen and aperture holders. These angles cause reduction of the stigmator strength, which can be compensated for up to about 70% by tilting the stigmator coils from their meridional position so that they are situated parallel to the deflection coils.

Since an attempt may be made to drive the objective lens in

the one-field-condenser mode, a superconducting condenser is installed near the objective. A second intermediate lens is added to permit a magnification higher than 10^6 with the two-stage normal projector included.

The performance of the system was tested. Preliminary results have been obtained with beam voltages of about 200 kV (Dietrich *et al.*, 1976).

After a starting time of 2 h the microscope can be used for 7 h without siphoning liquid helium into the inner helium chamber. Insertion of a holder with up to three specimens can be done with the lens system at liquid helium temperature. Positioning of each of the specimens in the electron beam is possible while the objective lens is excited.

Etching and contamination of the specimen is strikingly reduced. On the average the lateral specimen drift amounts to less than 0.1 nm/min. In one case a drift of less than 0.03 nm/min was observed.

In imaging an amorphous carbon foil, a resolution well below 0.2 nm could be obtained (Figure 5.10).

For an accurate determination of the resolving power, an evaluation of slightly displaced micrograph pairs was carried out

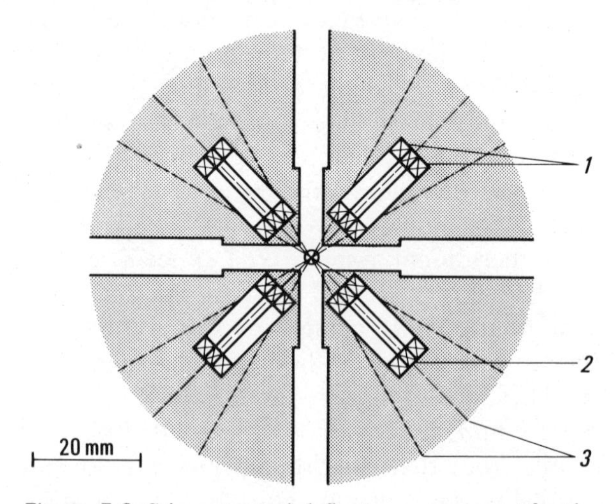

Figure 5.9. Stigmator and deflector arrangement for the system shown in Figure 5.7: (1) deflector coils; (2) meridional planes; (3) stigmator coils.

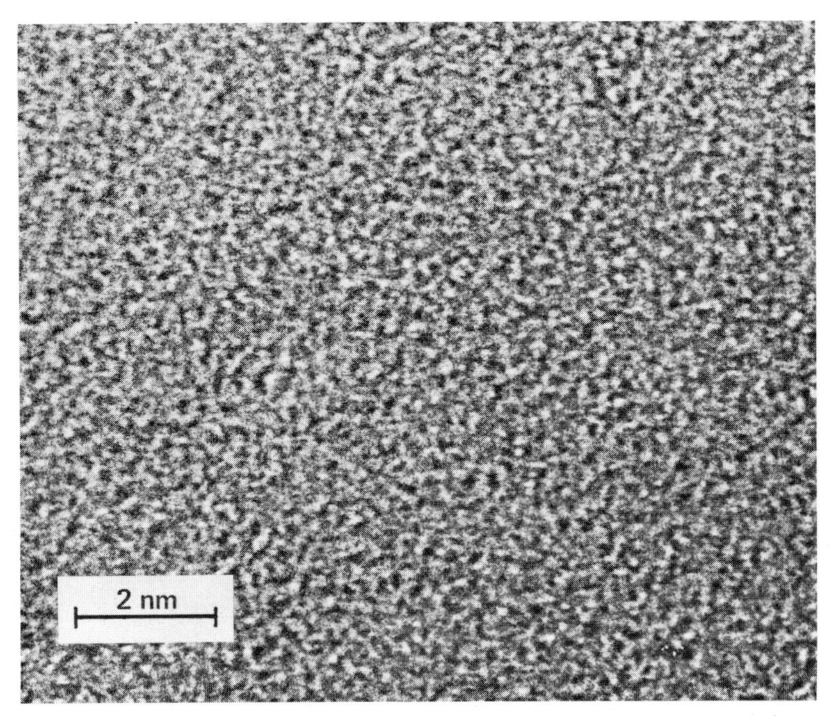

Figure 5.10. Carbon foil imaged with the system shown in Figure 5.7. (Micrograph taken by R. Weyl, H. Zerbst, and A. Feltynowski). Beam voltage 210 kV; beam current 4 nA; illumination aperture $\approx 10^{-4}$; objective lens data: $f_0 = 1.8$, $c_s = 1.45$, $c_c = 1.3$; defocusing distance $z_d = 40$ nm (overfocus); electron optical magnification 450,000.

(Hoppe *et al.*, 1970). Figure 5.11 shows the results (for different displacements of the pairs). The pattern modulation proves that the two micrographs are correlated, and this means that the same significant details exist on both images. The extent of the fringe system gives information on the maximum transferred space frequency. The different lengths of the fringes in Figures 11a and 11b are due to a change in the resolving power for the two displacement directions; this change may be caused by specimen drift or by a slightly tilted incidence of the electron beam.

The diffractograms shown in Figure 5.11 prove that it is possible to reach the theoretical limit of resolution in electron optical systems equipped with superconducting lenses.

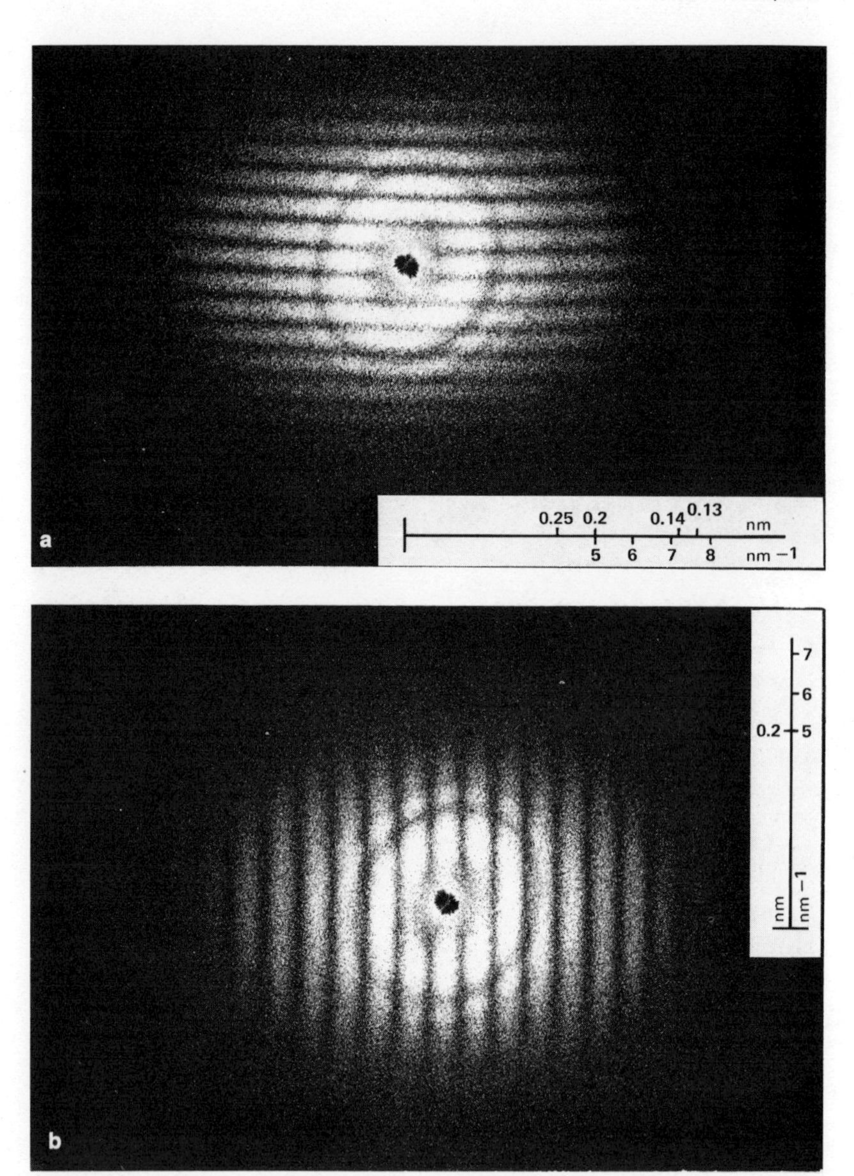

Figure 5.11. Light optical diffractograms of an image pair of micrographs, using that shown in Figure 5.10 and a somewhat displaced second micrograph taken immediately afterwards without changing the conditions. The displacement in (a) of the second micrograph is perpendicular to that in (b). (a) Resolution 0.14 nm; (b) resolution 0.18 nm.

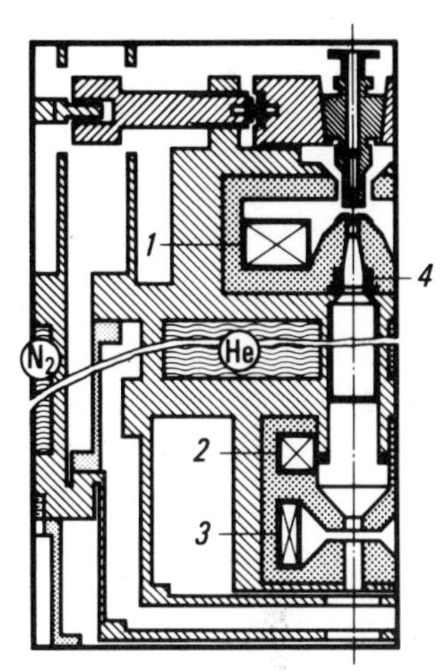

Figure 5.12. System of superconducting objective and two superconducting intermediate lenses. Reproduced by permission of Siegel *et al.* (1974). (▨) Liquid helium temperature (Cu); (▧) liquid nitrogen temperature; (▩) thermal insulator; (▧) sample holder; (1) objective lens; (2) low-magnification intermediate lens; (3) high-magnification intermediate lens; (4) stigmator.

5.2. Projected Systems

A system with superconducting lenses (Figure 5.12), developed and constructed by Siegel *et al.* (1974), will be incorporated into a high-resolution microscope for observing specimens under optimum conditions. The superconducting coils and iron circuits are clamped in pure copper blocks and cooled by thermal conduction. A 7-liter Dewar provides the helium supply. The iron circuit lens shown in Figure 5.12 can be exchanged for a superconducting ring or "disk" lens (lens A, Table 4.1). The cyrostat holds two intermediate lenses used to obtain different magnification ranges.

Proposals have been made for using superconducting devices in electron microscopes to investigate the possibilities of three-dimensional image reconstruction. For this purpose the specimen has to be imaged at different beam angles (Hoppe, 1974). In principle this could be done with a tilting stage, but the mechanical precision is not sufficient if atomic resolution is required. Hoppe is contemplating a system in which the beam instead of the specimen is tilted. Since tilt angles up to 45° are required, the spherical aberration in particular becomes large in radial directions. Special shapes for the lens fields, correction systems with a field stability of 10^{-8}, and the excellent vacuum conditions that can be attained with superconducting lenses are required.

6 Other Superconducting Elements for Electron Microscopy

6.1. Superconducting High-Voltage Beam Generator

The generator for high-voltage microscopes ($V > 1$ MV), in general a dc machine, is the largest single element of the instrument and causes its immense height. In Toulouse the height of the instrument is about 20 m (Dupouy and Perrier, 1972).

To reduce the size and still produce high-energy electrons, a microwave accelerator can be used, since it does not need a static voltage (Hart, 1966).

Normal-conducting linear accelerators with a length of the order of 1 m for 3-MeV electrons are already being used in medical and material research to obtain high-energy bremsstrahlung radiation. But in this case the requirements on energy spread and brightness of the radiation sources are moderate. These instruments can therefore not be used in electron microscopy. However, for superconducting microwave devices (Hart, 1966), the chances are better for achieving a small high-voltage electron ejector adequate for microscopes.

An electron linear accelerator (abbreviated LINAC, Figure 6.1) has the following characteristics. It consists of an electron emitter and resonators made of good conductors in which electromagnetic waves, normally standing waves, can be excited. An

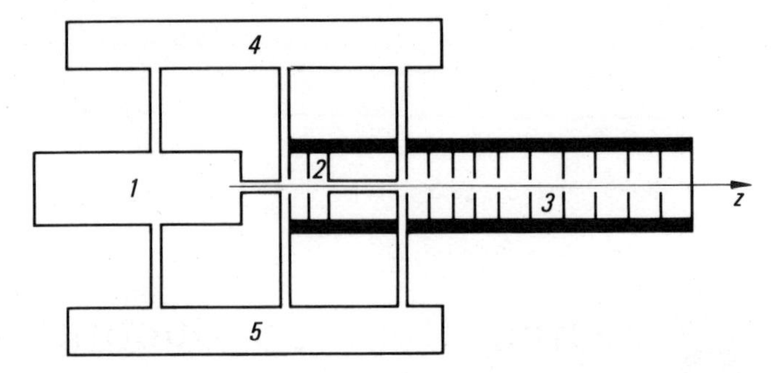

Figure 6.1. Diagram of a linear accelerator: (1) electron source; (2) buncher; (3) main cavity; (4) pumps; (5) high-frequency generator.

electron entering the accelerator in a phase in which it finds the electric field in the proper direction rides on the wave and is accelerated along the resonator if the phase velocity is matched to the increasing speed. This is illustrated in Figure 6.2. Optimum acceleration is achieved by those electrons for which the phase angle φ of the electric field **E** is zero at the entrance to the accelerator (i.e., at $l = l_0$). The electron energy at the end of the cavity is, for many cavity structures (e.g., the iris structures shown in Figure 6.3), proportional to the cosine of the phase at the entrance. For this reason the energy spread $\Delta E/E$ can only be kept

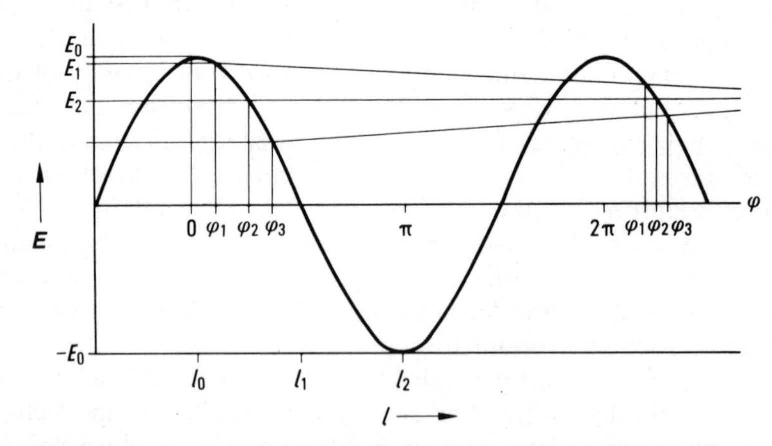

Figure 6.2. Electric field strength **E** along the accelerator at the time t_0. The diagram also illustrates the principle of phase focusing (see text).

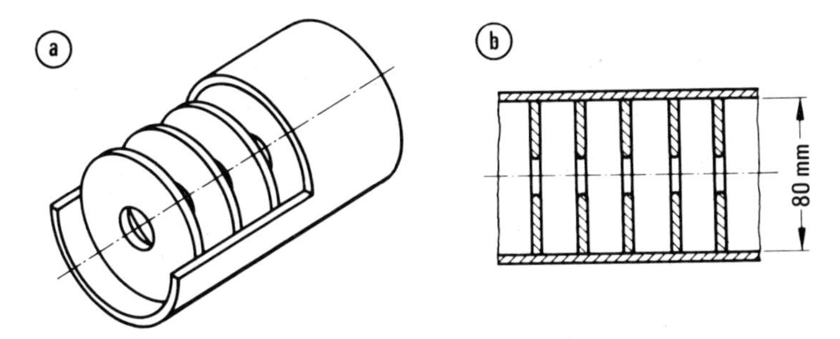

Figure 6.3. Cavity with iris structure. (a) Sectional view. (b) Cross section.

small if the phase interval $\Delta\varphi$, the bunch length, is small. Since electrons can be shot into the cavity only during the phase interval $\Delta\varphi$, the electron beam has a high-frequency structure.

The electrons are prebunched in the electron source. The short bunch length is obtained in the first resonator, called a buncher; here the velocity differences of the electrons, which are still nonrelativistic at this stage, may be utilized for phase focusing. The principle of phase focusing can be understood from Figure 6.2. The resonator is constructed so that the velocity of the high-frequency wave equals the velocity of the electrons at φ_2, and so the wave and the electron arrive at φ_2' at the same time ($\varphi_2' - \varphi_2 = 2\pi$). Electrons reaching the buncher earlier meet the smaller field strength E_3 at φ_3, are thus not so strongly accelerated, and $\varphi_3' - \varphi_3 < 2\pi$; electrons which reach the buncher later have a larger acceleration, so $\varphi_1' - \varphi_1 > 2\pi$. This results in a decrease of the bunch length from $\Delta\varphi = \varphi_3 - \varphi_1$ to $\Delta\varphi = \varphi_3' - \varphi_1'$ after one wavelength. In this way the bunch length can be reduced considerably. For small variations in the energy spread the permissible overall bunch length has to be of the order 1%, which means the microduty cycle is small. Further acceleration takes place in the main resonator.

The main problem in the application of a LINAC in electron microscopy is achieving an electron beam with a high brightness, on the order of at least 10^5 A cm^{-2} sr^{-1}, combined with a small energy spread, i.e., $(\Delta E)/E \leq 10^{-5}$. One has to obtain a high rate of energy gain in order to keep the accelerator short in length, and this means strong electric fields and large currents in the resonator wall. In normal LINACs the losses in the walls prohibit continuous operation of the instrument since adequate cooling is not possible. Thus the LINACs are operated on a small duty cycle, i.e., they are pulsed only a fraction of a percent of the time. The pulses are superimposed on the high-frequency structure. This has many disadvantages, especially in regard to the brightness, considering the already small microduty cycle. Continuous performance is possible only by using cavities made of superconducting material with low surface resistance and with quality factors five orders of magnitude higher, as discussed in Section 3.2. For this reason superconducting LINACs were proposed for electron microscopes with standard resolution. Such superconducting devices with Pb cavities are being studied at Osaka University (K. Ura, private communication). Details of the construction of a superconducting LINAC in a 3-MV instrument are given in Chapter 7.

The optical properties of LINACs have been investigated by several authors (Boussard, 1967). Their use in microscopes has been studied mainly by Matsuda and Ura (1974).

The accelerator is in effect a long lens with a continuously changing index of refraction. It can be represented by a number of thin lenses which together give a defocusing effect. This defocusing force f_r is exerted on the electrons by the radial electric field component E_r:

$$\begin{aligned} f_r &= \frac{d}{dt}(mv_r) \\ &= eE_r \\ &= \frac{eE_0}{2} r \frac{2\pi}{\lambda_{\mathrm{HF}}} \frac{1-\beta^2}{\beta} \sin\varphi \qquad (6.1) \end{aligned}$$

Here v_r is the radial velocity, $\beta = v_z/c$, v_z is the axial velocity, r is the distance from the z axis, λ_{HF} is the high-frequency wavelength, $E_0 = |\mathbf{E}_0|$ is the amplitude of the electric field (Figure 6.2). From relation (6.1) it follows that the diverging force depends on the velocity and phase of the individual particle. It is larger the smaller β is, i.e., the main defocusing effect takes place near the entrance of the buncher. The calculations make use of a time- and momentum-dependent transformation matrix which connects the position and momentum of the particles at the accelerator entrance with the corresponding values at the exit (Passow, unpublished). The matrix presentation for calculating the imaging property of lens systems is described by, e.g., Septier (1967).

As pointed out by Scherzer (1947), the defocusing property of the accelerator can be used to build an achromatic system analogous to light optical systems where diverging and converging lenses are combined for this purpose. An example is given in Chapter 7.

6.2. Magnetic Dipoles

Magnetic dipoles have already been mentioned as correction systems for rotationally symmetric lenses where it is necessary to compensate for deflection angles well below 1°. The main problem in this case is how to install correction systems with the small available space. Dipoles may also be used for changing the beam direction in the microscope, thus introducing a bend in the microscope layout (von Zuylen and Fontija, 1974). Besides the separation of electrons with different energies, such a spectrometer offers further possibilities for reducing the energy spread. Haine and Jervis (1955), for example, described a spectrometer for stabilizing a 50-kV beam.

Different kinds of spectrometer magnets, including the results of calculations to correct for the edge effect, are discussed by Livingood (1969). A rough sketch of one type, a window-frame-sector type, is shown in Figure 6.4. The somewhat problematic fringing field effects are indicated in Figure 6.4c. This magnet can, for example, deflect a beam emerging from a horizontal accelerator and entering a vertically constructed column. The focusing effect due to the different lengths of the paths in the magnet for different trajectories is indicated in Figure 6.4b. A spectrometer is included in the proposed 3-MV superconducting microscope described in the next chapter.

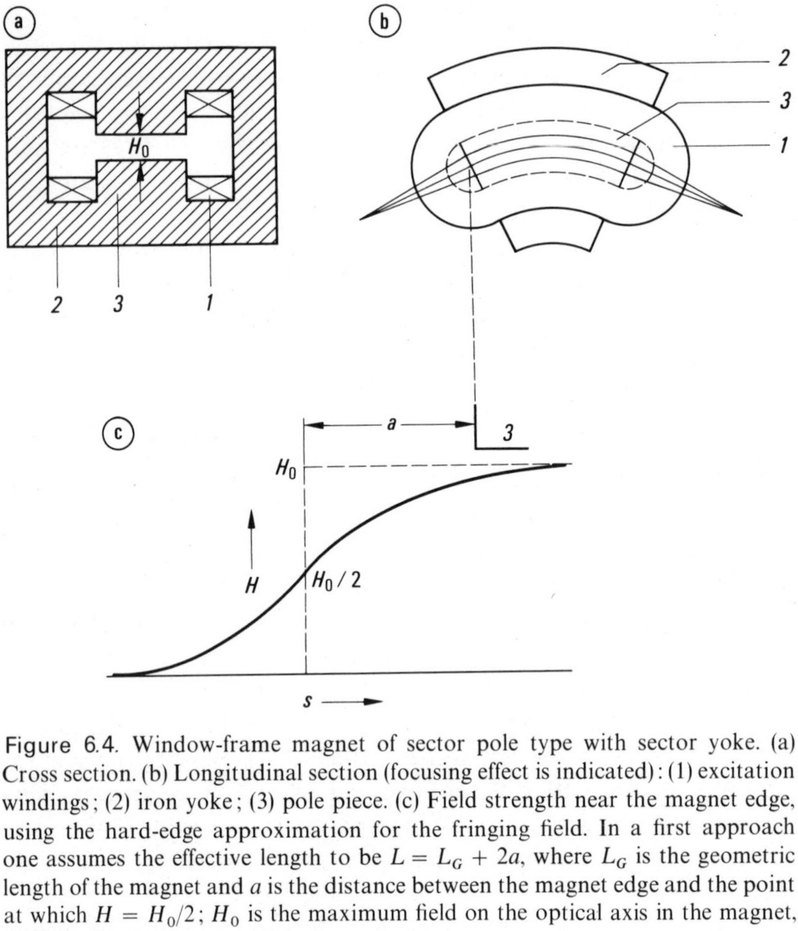

Figure 6.4. Window-frame magnet of sector pole type with sector yoke. (a) Cross section. (b) Longitudinal section (focusing effect is indicated): (1) excitation windings; (2) iron yoke; (3) pole piece. (c) Field strength near the magnet edge, using the hard-edge approximation for the fringing field. In a first approach one assumes the effective length to be $L = L_G + 2a$, where L_G is the geometric length of the magnet and a is the distance between the magnet edge and the point at which $H = H_0/2$; H_0 is the maximum field on the optical axis in the magnet, and s is the path length along the optical axis.

As the spectrometer field strength for applications in electron microscopy should not exceed $\mu_0 H \approx 2\text{T}$ for beam voltages up to 5 MV, iron pole pieces might be the rule. Nevertheless, the superconducting version is interesting for obtaining the required field stability, which should be of the order of $\Delta H/H = 10^{-7}$ if the desired energy spread is $\Delta E/E \leq 10^{-6}$. In addition, the excellent cryogenic vacuum is very useful in many cases.

7 Proposed Superconducting 3-MV Microscope

A possible construction for a 3-MV microscope with most components of the superconducting variety and operated mainly in the fixed-beam mode is outlined in the literature (Dietrich *et al.*, 1975). Passow (1976) has considered means for improving the resolution, especially for the case when the microscope, which has a microwave accelerator, is operated in the scanning mode.

The high-voltage microscope sketched in Figure 7.1 will be about 1.5 m high. The superconducting microwave beam generator consisting of the high-frequency emission source, the buncher, and the main accelerator is connected to the lens system by a superconducting 180° spectrometer. This is used not only for deflecting the beam but also for controlling the electron energy by means of position detectors. It may also serve for energy selection.

The image observation and recording units are envisaged as adjuncts in conventional design. All superconducting components should be cooled with superfluid helium, not only to reduce vibrations, but also to dissipate the heat more quickly—essential for the performance of the microwave resonator.

7.1. Accelerator

The goal of the microwave generator design is to combine an increase in brightness with a reduced energy spread. For this reason the construction shown in Figure 7.2 is more complicated than that of the Stanford superconducting LINAC for high-energy electrons (Suelzle, 1973); parts of the Stanford accelerator have already been tested. In general, the energy spread for the beam depends mainly on the electrons traveling in the accelerator field with different phases, as pointed out in the last chapter. For a short phase length, the microduty cycle has to be kept small, and this results in a decreased mean brightness.

However, sufficient brightness can be achieved by increasing the current density of the field emission source. An intensity increase is possible with the proposed microwave field emission source. It consists of a cavity with its resonance frequency at about $\nu_0 = 24$ GHz. The electrons are drawn out of the tip by the electric component of the microwave field. In comparison to dc sources such as thermionic or field-emission cathodes (Crewe *et al.*, 1968; Veneklasen, 1972) that are used in conventional microscopes, the pulsed gun has several advantages, such as

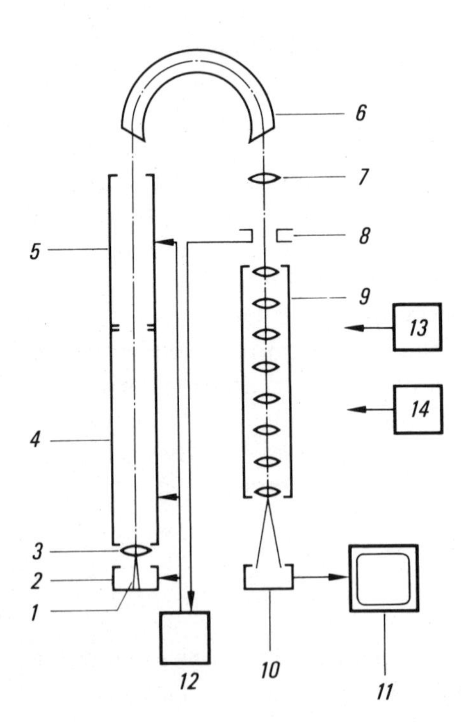

Figure 7.1. Diagram of the 3-MV microscope (Dietrich *et al.*, 1975): (1) field emission source; (2) source cavity; (3) adapting lens and electrostatic pre-accelerator; (4) buncher; (5) main accelerator; (6) spectrometer; (7) adapting lens; (8) beam position detector; (9) lens system; (10) image intensifier system; (11) monitor; (12) HF generator with regulation units; (13) electronics for lenses, spectrometer,and correction system; (14) helium liquefier.

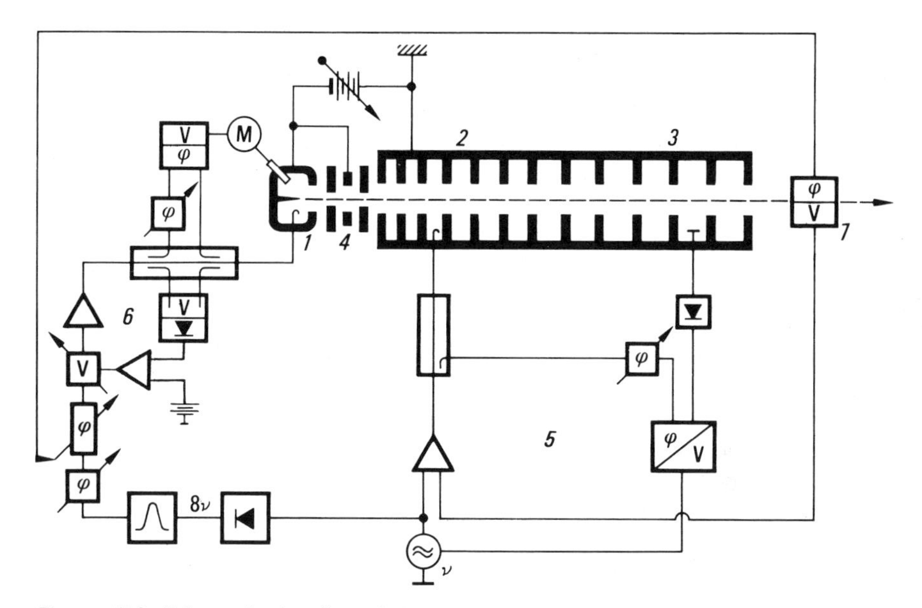

Figure 7.2. Schematic drawing of the accelerator system with source and the control system for the 3-MV microscope (Dietrich *et al.*, 1975): (1) source; (2) buncher; (3) accelerator; (4) gun (the source, buncher, accelerator, and gun are combined in one cavity); (5) control system of the cavity voltage and frequency; (6) control system of the source phase and voltage; (7) beam detector. φ is the phase, V the voltage, ν the frequency, and M a motor.

higher emission current during the small microwave cycle time and better stability and longer lifetime of the cathode as a consequence of the smaller ion impact. The latter can be explained by the fact that the average traveling length of the ions (in this case mainly He ions) per microwave cycle will be small compared to the distance from the tip to the anode.

The field emission current I can be calculated using the Fowler–Nordheim equation, which when modified for this case is

$$I = I(\mathbf{E}_Q)\exp(-\varphi/\varphi_Q)^2 \tag{7.1}$$

where the subscript Q is used to designate values for the electron source; the field strength in the cathode is given by

$$\mathbf{E} = \mathbf{E}_Q \cos\varphi$$

and the time-dependent phase is given by

$$\varphi = 2\pi\nu_Q t$$

Here ν_Q is the operation frequency, and φ_Q is the width of the distribution function and is calculated to be $\approx 20°$. The electron energy E_Q for phase φ on ejection of the electrons from the source amounts to

$$E_Q = E_{Q,\varphi=0} \cos\varphi \pm \Delta E_F \tag{7.2}$$

Here ΔE_F is the energy spread of a dc field emission gun and is ≤ 0.5 eV.

The energy deviation due to the phase deviation should be 6% according to equation (7.1). The energy distribution is sketched in Figure 7.3.

The accelerating system is designed to increase the electron energy and to reduce the energy spread. An electrostatic preaccelerator with a dc voltage of $\approx$ 15 kV is applied so that the electrons gain a certain velocity; this makes the requirements on the buncher less stringent. The buncher has to reduce the phase length, as mentioned in Chapter 6, and to accelerate the electrons to relativistic energy (0.58 MeV). In this system, however, its additional and even more important task is to rotate the electron distribution in phase space. The buncher as well as the accelerator are operated at a frequency of 3 GHz, and with this trick the bunch length of 20° at injection in the buncher is reduced to 5°. However, seven of eight bunches of the microwave source are lost.

The buncher length is chosen so that the electrons will be ejected from the buncher when the distribution pattern is rotated by an angle of 180°. Particles with a higher energy will travel faster and therefore gain phase. They will run into areas with lower field strength and consequently gain less energy, and, *vice versa*, particles initially with less energy will gain more. This is why the energy spread in phase space increases by the same factor (bunching factor) by which the phase spread is reduced (Figure 7.3).

In the main accelerator the phase of the accelerating wave is shifted relative to the wave in the buncher, so that the electrons ride on the crest of the microwave to gain the final energy of 3 MeV. The particles with the energy for which the accelerator is

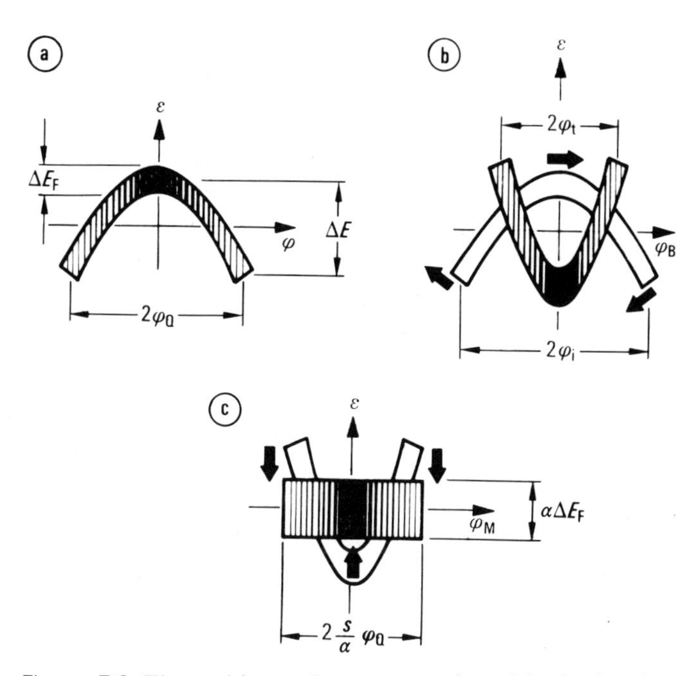

Figure 7.3. The position and movement of particles in the phase space of the accelerator (Dietrich *et al.*, 1975). The coordinates are the relative energy deviation ε of the particles from the synchronous particle and the phase difference φ between the synchronous particle (index S) and the real particles, i.e., the origin is defined by the synchronous particle. (a) Particle distribution upon emission from the source; the time dependence of the phase is given by $\varphi = 2\pi\nu_Q t$; ΔE is the energy spread at the source and φ_Q is the phase spread at the source. (b) Distribution at injection (unshaded strip) and ejection (shaded strip) in the buncher (index B); the phase φ_{SB} of the synchronous particle is not equal to zero; the particles are bunched, rotated, and accelerated; $\varphi_i = s\varphi_Q$, where s is the stretching factor, is the phase spread at injection into the buncher; $\varphi_t = 1/\alpha\varphi_i$, where α is the bunching factor, is the phase spread at ejection from the buncher. (c) Distribution at injection (unshaded strip) and ejection (shaded strip) in the main accelerator (index M); the phase φ_{SM} of the synchronous particle is equal to zero; the particles are accelerated and stretched out. The time dependence of the phase in (b) and (c) is given by $\varphi = 2\pi\nu t$, with $\nu_Q/\nu \approx 8$.

designed called synchronous particles since they arrive at the right time are accelerated with a phase $\varphi_S = 0$.

Particles with zero phase, but in a too-low energy state, will travel in a higher field and gain more energy than those which are injected with phase deviation but with higher energy. Thc accelerating process will be finished when the distribution has a rectangular form. For electrons which travel with the same velocity, the energy spread is then equal to the energy spread ΔE_F of the dc gun multiplied by the bunching factor. The deviation of the electron velocity from the velocity of light leads to an additional energy spread, although this spread is small if the field strength in the main accelerator is high, as in the machine described here.

Passow (1976) has shown that if the conditions briefly sketched here have to be fulfilled, there are still several degrees of freedom left so that a practical design with a relative energy spread of $\Delta E/E \leq 10^{-5}$ should be possible. The data are given in Table 7.1.

Table 7.1. Parameters of the 3-MV Microscope Accelerator

Final energy	3 MeV
Design value of the energy spread $\Delta E/E$	1×10^{-5}
Energy spread of the ideal accelerator system $\Delta E/E$	3×10^{-6}
Ejection energy of the source	2 keV
Energy gain in the	
preaccelerator	15 keV
buncher	0.58 MeV
main accelerator	2.42 MeV
Operation frequency of the	
source	24 GHz
accelerator system	3 GHz
Total length of the accelerator system	1.90 m
Length of the	
buncher	1.00 m
accelerator	0.81 m
Maximum accelerating field strength in the accelerator system	3 MV/m
Field emission tip	
field strength	5×10^9 V/m
current density	6×10^9 A/m^2
Peak current	57 μA
Average current	6.4 μA
Brightness of the accelerator system without correction lens	3.3×10^7 (A cm^{-2} sr^{-1})

A further decrease of the energy spread by using apertures at the exit of the spectrometer or by using more lenses to obtain an achromatic effect are therefore not planned at present.

The brightness in this system should be at least equal to that in a conventional machine of equal voltage with a thermionic cathode. The brightness R is defined as

$$R = \frac{\partial^2 I}{\partial F \partial \Omega} = \frac{\partial^2 I}{\partial A_x \partial A_y}$$

where F is the emission area, Ω the solid angle, and A the phase space area. The total phase space area filled by all the electrons accelerated up to the final energy was calculated considering their energy and phase deviation and the chromatic error of the accelerator. Though the average brightness of a microwave accelerator is reduced by a factor

$$n = \frac{dI_{\mathrm{ac}}}{dI_{\mathrm{dc}}}$$

(Rymer, 1966), a sufficiently large value for the brightness can be expected in this system, first because a field emission source is used and second because the bunch length is considerably larger than in machines described elsewhere. In addition it might be possible to draw a higher maximum current out of a pulsed source than out of a dc source. Thus an R value of $\approx 5 \times 10^7$ A cm^{-2} sr^{-1} should be a reasonable estimate.

To ensure a stable run of the machine, control systems with feedbacks have to be provided as shown in Figure 7.2. The beam position detector behind the spectrometer produces two signals, a signal which displays a shift of the central energy, and another one which depends on the beam diameter. They control the power input into the cavity and in this way control its maximum field strength and the phase between the source cavity and the main cavity. To obtain a frequency of ≈ 24 GHz in the source cavity, the resonance frequency of the main cavity is increased by a frequency multiplier. The tuning takes place mechanically. The field strength will be controlled by comparing a rectified signal with a dc voltage.

7.2. Spectrometer

The 180° magnetic spectrometer shown in Figure 7.4 is designed for a radius of 0.3 m. Due to the extension of the cryostats, the accelerator and the lens system are separated by a distance of 0.6 m. A picture-frame magnet with sector pole and sector yoke is chosen. The superconducting version is of advantage mainly because of the vacuum requirements. The field strength of the spectrometer can be calculated from the equation

$$H = \frac{1}{\mu_0 rc}(E^2 + 2E \cdot E_0)^{1/2}$$

where the kinetic energy E of the electrons is 3 MeV, their rest mass energy E_0 is 0.511 MeV, and c is the speed of light.

A distortion-free transfer of the crossover at the accelerator exit into its image behind the spectrometer is desired to avoid additional correction elements. The radial displacement of the electrons in the image due to the aperture angle α can be compensated for by a radial dependence of the field,

$$\mu_0 H \propto 1/r^n, \qquad n = -\frac{r}{H}\frac{dH}{dr}$$

which produces radial oscillations of the particles causing a focusing effect.

The axial focusing can be achieved by supplying the magnet with appropriate edge angles, since the fringing fields at the edges of the magnet (Figure 6.4) introduce radial components which act as thin radically defocusing lenses. The quantitative relations are summarized by Livingood (1969). The data of this spectrometer, which is a symmetric type with a 1:1 imaging scale, are given in Table 7.2. By means of the spectrometer, the energy of the electrons emitted from the accelerator is controlled, as mentioned above, with an accuracy of $\Delta E/E = 10^{-5}$, i.e., the crossover which is imaged by the spectrometer must have an extension $\Delta x \leq D(\Delta E)/E = 14\ \mu\text{m}$ if D is the energy dispersion of the instrument. The electron detector for signaling the energy should be placed in the plane of the crossover image behind the spectrometer. Since the

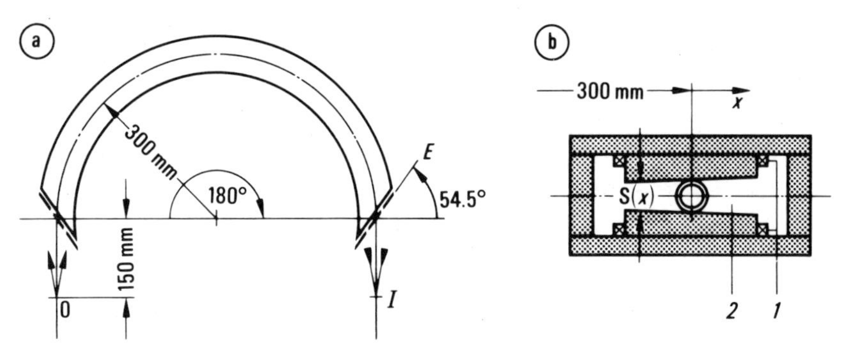

Figure 7.4. 3-MV microscope spectrometer (Dietrich *et al.*, 1975). (a) Shape of the pole piece: (O) objective plane; (I) image plane; (E) effective edge. (b) Cross section: (1) superconducting coils; (2) vacuum tube. The gap separation S is given by $S(x) = 10(1 + x/300^{0.286})$ in mm.

detection of a beam deviation $\Delta x \approx 10\ \mu\text{m}$ from the desired value is a strong demand, a one-stage magnification is provided. A fixed-focus lens with a short focal length is appropriate, and this led to the choice of the shielding lens (Figure 4.13) for this purpose.

Since the high-frequency structure of the beam allows use of beam detectors based on the induction principle, it is not necessary to touch the beam. Bergere *et al.* (1962) described a system of four short induction loops arranged around the crossover image parallel to the beam. When connected to a difference amplifier, the loops opposite each other produce a signal proportional to

Table 7.2. 3-MV Microscope Spectrometer Data

Spectrometer angle	180
Beam path radius	0.3 m
Gap induction	38.6 mT
Average gap width	10 mm
Excitation	308 A
Imaging scale	−1
Field index n	0.286
Object distance and image distance from the effective field edge	0.15 m
Edge angles	54.5°
Distance of the effective field edge from the pole-piece edge	10.4 mm
Dispersion	1.43 m
Tolerable source diameter for energy resolution $\Delta E/E = 10^{-6}$	1.4 μm

the deviation from the desired position. If the distance between the induction loops and the beam is about 1 mm, a deviation of the order of the beam diameter should be detectable. If the detection system still lacks enough sensitivity, either superconducting cavities for magnifying the inductive voltage or Josephson junction devices such as SQUID (*s*uperconducting *qu*antum *i*nterference *d*evices) with a voltage sensitivity of the order of 10^{-14} V can be used.

7.3. Microscope Column

For optimum design, the column should be compact, some of the parts should be interchangeable, and there should be good accessibility to all parts. The number of mechanical adjustment devices has to be restricted as far as possible. To achieve the desired small illuminated specimen spot and the high magnification of 10^6 without enlarging the column length far beyond 1 m, a larger number of lenses in the projector and condenser system is proposed than is used in conventional microscopes (Figure 7.5). Separate cryostats (Figure 7.6) are provided for the three-stage condenser, the objective lens with intermediate lens, and the three-stage projector. In order to cool the system down starting from regions far away from the beam, the cryostats consist of two interconnected helium chambers with the helium inlet in the outer chamber.

The construction of the condenser and projector system is shown in Figure 7.7. Ferromagnetic circuit lenses are used. The coils and the iron circuits are immersed in liquid helium and fastened to the inner tube (vacuum tube) of the cryostat. For easy substitution of defective parts, the pole pieces, correction systems, and apertures are installed in a holder tube inserted in the vacuum tube. A slightly modified type H lens (Figure 4.8) is considered appropriate for this purpose since the field distribution is at least partially determined by the surrounding iron. Fastening the pole pieces in the holder tube will favor the development of secondary lenses, but this can be prevented by using jackets of a strongly

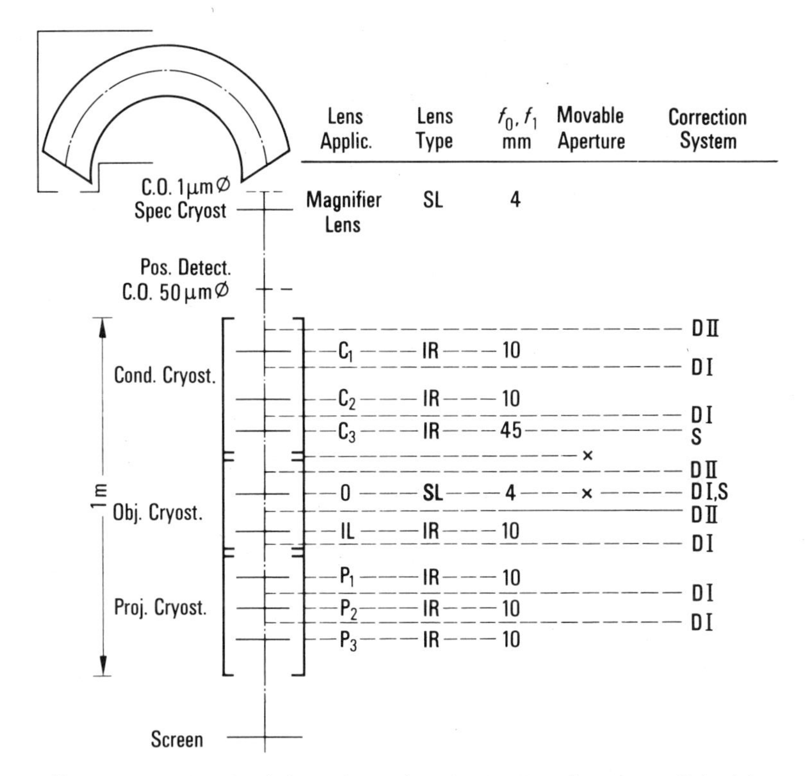

Figure 7.5. Sketch of the column for the 3-MV microscope (Dietrich *et al.*, 1975): (DI) one-stage deflection system; (DII) two-stage deflection system; (S) stigmator; (f_0, f_1) focal lengths; (IR) iron core lens; (C_1, C_2, C_3) condenser lenses; (O) objective lens; (IL) intermediate lens; (P_1, P_2, P_3) projector lenses; (SL) shielding lens; (C.O.) crossover.

shielding superconductor, e.g., Nb–Sn sinter material as described in Section 3.2.

There will be at least two different objective cryostats equipped with the same intermediate lens. For high resolution, the shielding lens with its short focal length and excellent possibilities for compensation of field errors is especially suitable. However, the investigation of thicker samples at different temperatures requires a larger gap. For this purpose a low-field, large-focal-length lens has to be installed, although the total magnification will be reduced.

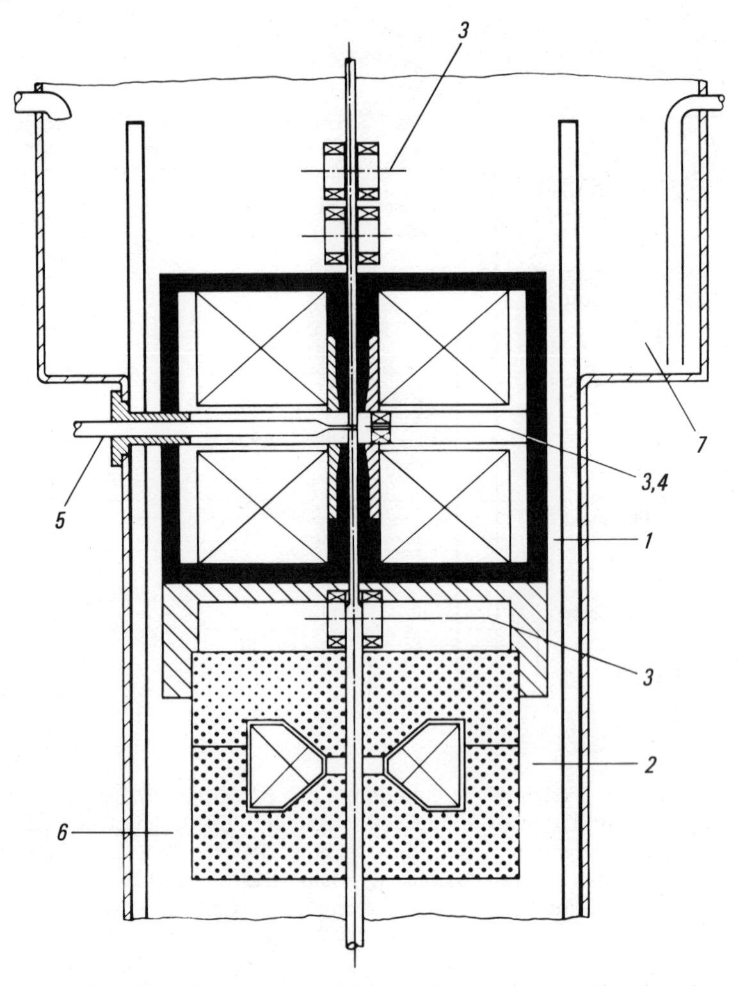

Figure 7.6. Objective cryostat with objective and intermediate lens for the 3-MV microscope (Dietrich *et al.*, 1975): (1) objective lens; (2) intermediate lens; (3) deflection systems; (4) stigmator; (5) side-entry for specimen; (6) inner liquid helium chamber; (7) outer liquid helium chamber; (■) Nb_3Sn; (▩) Fe.

Mechanical shifting devices for adjustments can be replaced by electromagnetic deflection systems as pointed out before. In the condenser and the projector system, a simple folding of the beam near the lens planes is sufficient to harmonize the center of rotation of the single lenses. However, a careful adjustment of the

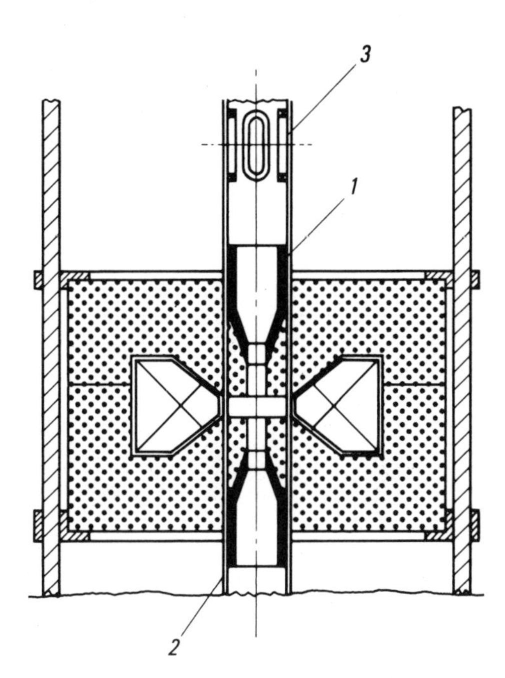

Figure 7.7. Condenser and projector lens for the 3-MV microscope (Dietrich *et al.*, 1975): (1) superconducting jacket; (2) holder tube; (3) one-stage deflector; (■) Nb_3Sn; (▩) Fe.

beam along the magnetic axis of the objective is necessary. For this purpose a two-stage deflecting system is provided; this system can also be used to shift the beam if the microscope is used in the scanning mode. Since good alignment of the objective and intermediate lens axes is decisive, a two-stage deflector also has to be put in behind the objective lens. Stigmators are necessary for the third condenser stage, for the objective, and eventually for the intermediate lens.

The deflector and stigmators fastened in the holder tube are made of flat superconducting coils wound from thin superconducting wires with a high critical temperature and, if possible, intrinsically stabilized, since they are not surrounded by liquid helium and are cooled only by thermal conduction (Figure 7.7).

The characteristic data for the lenses to be considered here are summarized in Table 7.3. The condenser system (C_1, C_2, C_3) must reduce the crossover of the beam in the plane of the position detector by a factor of 50 in order to produce an illuminated spot 1 μm in diameter. Since C_1 and C_2 are of small focal length, they

Table 7.3. Characteristic Lens Data for the 3-MV Microscope

Quantity[a]	Lens			
	C_1, C_2,[b]	C_3	O[c]	IL, P_1, P_2, P_3[b]
B_0, T	5	0.5	6	5
NI, kA	66	20	—	66
$2s$, mm	6	50	—	6
$2r$, mm	3.5	20	7	3.5
d, mm	5.5	30	1.3	5.5
k^2	1.5	0.8	1.3	1.5
f_0, mm	6.25	45	3.8	6.25
f_1, mm	10.5	45	3.8	10.5

[a] B_0 is the maximum induction on the optical axis, NI the excitation, $2s$ the gap width, $2r$ the bore diameter, $2d$ the half-width, k^2 the lens strength, f_0 the focal length (objective mode), f_1 the asymptotic focal length (projector mode).
[b] Lens H, Table 4.2, slightly modified.
[c] Data taken from numerical calculations similar to those of Dietrich *et al.* (1972*a*).

cause a considerable reduction. Lens C_3, however, controls the illumination and is equipped with a wide bore hole, so its focal length is large in spite of a high excitation. It magnifies by a factor of 3 at the highest excitation. Movable and interchangeable apertures are provided between C_3 and the objective lens O to vary the aperture angle.

The illumination data are dependent on the prefield of the objective lens. If the objective is operated in the one-field-condenser mode, the aperture is imaged into the specimen by the prefield with a reduction factor of 100, and the illumination is determined by the size of the crossover in the entrance pupil. The latter mode yields a very small illuminated spot which is useful for scanning applications.

The objective lens for high resolution has to be tightly connected with the intermediate lens (Figure 7.6), as was found necessary by tests with the system described in Chapter 5 (Figure 5.4).

The whole arrangement is similar to the system shown in Figure 5.7, but the dimensions are slightly changed for the higher voltage. Most of the arguments discussed in connection with the system of Figure 5.7 are also valid here. Since the gap between the cylinders is 7 mm, there is sufficient space for a side entry, apertures, and correction systems as sketched in Figure 5.7.

Under normal conditions apart from the scanning, a lens strength $1 < k^2 < 2$ would be suitable. Because of the short wavelength, the theoretical resolution is kept well below 0.1 nm with the data listed in Table 7.3. The magnification of the objective and intermediate lens is approximately 1000-fold. The reduction in magnification for the case of a long-focal-length objective lens with a gap width of 20–30 mm would be by a factor of about 3.

The three projector lenses P_1, P_2, P_3 permit a maximum additional magnification of 1000. Lower magnification can be achieved by switching off one or two of the projector lenses. With the shielding lens as the objective and only P_3 operating, the minimum overall magnification would be 5000.

For operating the instrument in the scanning mode the intermediate lens and the projector are, as a rule, shut down. They are used only for adjusting the illuminated spot.

For electron diffraction with a maximum camera length of about 4 m, the diffraction pattern produced is transferred from the objective lens to the objective plane of P_1 by means of the intermediate lens. Smaller diffraction lengths can be obtained by switching off projector lenses. The image observation system is constructed in the conventional way without any superconducting parts and therefore the television intensifier is not described here.

The resolution of the microscope was determined by cal-

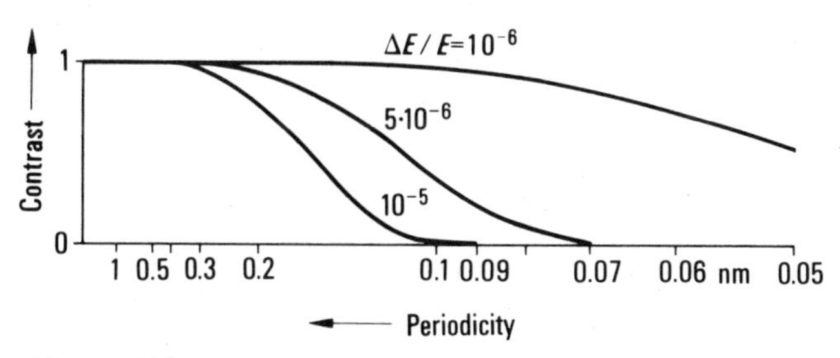

Figure 7.8. Envelope of the transfer function for various energy width $\Delta E/E$ (Dietrich *et al.*, 1975). The resolution is limited by vanishing contrast. The electron energy E is 3 MeV, the spherical aberration coefficient c_s is 3.5 mm, and the chromatic aberration coefficient c_c is 3 mm.

culating the envelope of the transfer function (Hanssen and Trepte, 1971) in the relativistic case, taking into account the data for the accelerator ($\Delta E/E = 10^{-5}$) and aberration coefficients of the shielding lens ($c_s = 3.5$, $c_c = 3.0$). It can be seen from Figure 7.8 that the energy width $\Delta E/E = 10^{-5}$ impairs the transfer function only below a periodicity of ≈ 0.3 nm. For further improvement in the resolution, the energy spread of the accelerator has to be reduced as envisaged in the following sections, or else energy selection must be used.

7.4. Further Improvements of the System

The microwave accelerator may be improved by employing certain principles outlined by Passow (1976); these principles should be especially useful for high-voltage scanning microscopes.

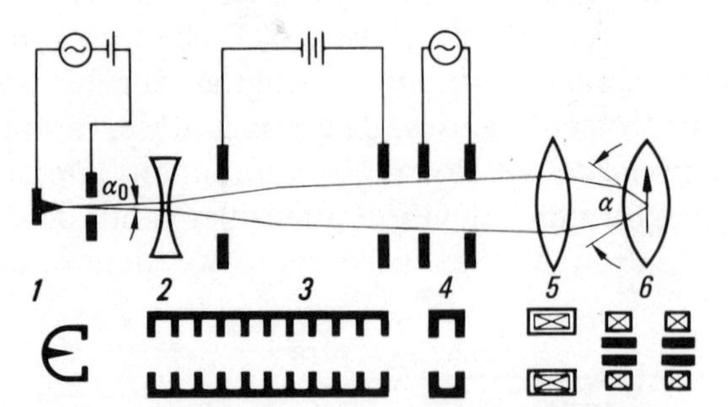

Figure 7.9. Microscope with superconducting LINAC—the improved system in the scanning mode. Reproduced by permission of Passow (1976). (1) Electron source; (2) first part of accelerator and symbol for its defocusing effect; (3) second part of accelerator and symbolic representation of its accelerating effect; (4) microwave condenser, which causes a phase-dependent change in the energy distribution; (5) dc condenser; (6) dc condenser objective, a superconducting shielding lens; the specimen is mounted in the lens center. α_0 and α are aperture angles.

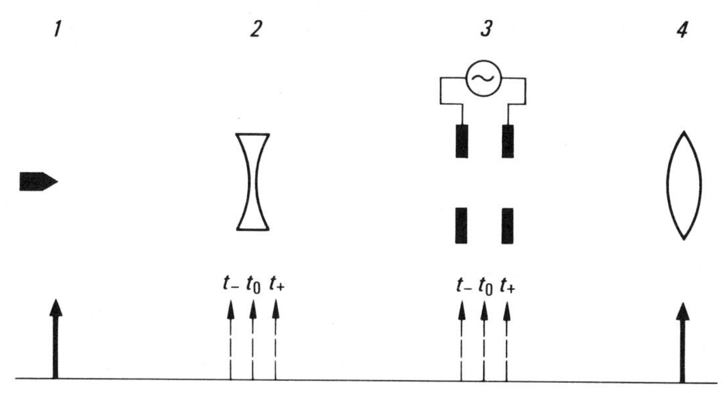

Figure 7.10. Schematic representation of the compensation of time-dependent errors due to the accelerating system by the static chromatic error of the condenser field. Reproduced by permission of Passow (1976). (1) Image of the source. (2) Virtual image produced by the diverging lens, which represents the accelerating system. The images formed by particles arriving at different times $t_+ = t_0 + t$ and $t_- = t_0 - t$ are partially separated. The energy, however, is identical in the approximation assumed here. (3) Virtual image after the energy modulation. The relative location of the images does not change; the energy distribution is $E(t_+) = E(t_0) + E$, $E(t_-) = E(t_0) - E$. (4) The modulation of the energy E results in an E chosen so that the chromatic error of the reducing lens shifts all images to the same point.

According to the theoretical work (Scherzer, 1947), the chromatic and spherical aberration in electromagnetic lenses can be corrected if time-dependent fields are used. With the proposed procedure it is not completely possible to compensate for the chromatic aberration caused by the energy distribution of the electron source, the ripple of the lens current, or the scattering in the specimen. But the deviations resulting from periodic oscillations of the energy distribution, which would result in a voltage stability $\Delta V/V = 10^{-5}$, can be corrected.

Figure 7.9 shows a possible arrangement of the parts of the improved system and symbols for the optically relevant parts. Since an accelerator is in effect a diverging lens with a focal length that increases as the electron energy increases (Chapter 6), the symbol representing this lens effect is located at the beginning of the accelerating system, where the electron energy is still low. The buncher and the part of the main accelerator in which the electrons are accelerated but not deflected are symbolized by a dc voltage. The end of the main accelerator is used to change the energy dependence of the phase. This is performed in such a way that the phase-dependent focal length of the diverging lens modulated by the phase-dependent energy yields a negative, time-independent chromatic aberration, as indicated in Figure 7.10. If the microscope is operated in the scanning transmission mode, the magnitude of the resulting chromatic aberration is chosen so that it compensates for the positive chromatic aberration of the condenser and for the effect of the objective lens on the beam before it reaches the specimen (Figure 7.10). This results in an extremely small illuminated spot on the specimen if the other optical parameters are appropriately chosen.

If such a correction is desired for a fixed-beam instrument, another lens is necessary. With the right arrangement, the phase spread is diminished by this lens, and the remaining energy spread is compensated for by an additional cavity.,

There is no question that the corrections described can be performed only with superconducting devices. Otherwise the spatial extension of the additional components would make the microscope susceptible to mechanical vibration, which could be fatal for the desired high-resolution images.

* | Appendixes

A. Superconducting Electron Optical Systems for High-Energy Physics

A.1. General Remarks

Aside from superconducting microwave accelerators, the principle of which has already been outlined in connection with electron microscopy, the possible applications of superconducting electron optical devices in high-energy physics are mainly as magnets. However, for the electron accelerators and storage rings now existing and under study, the superconducting-magnet technology is not as essential as it is for some of the heavy-particle machines. This is probably one reason that systems with superconducting magnets are still in the project phase.

In electron microscopy, rotationally symmetric lens systems are the most important components. However, focusing the electrons in the GeV region is often better done by quadrupole lenses, especially in the case of a large aperture angle. Because of the small bending angles and the relatively wide energy spread, the requirements for dipole magnets included in beam guides differ in some respects from the requirements for the spectrometer described in Chapter 5. In high-energy physics, pulsed magnets are frequently used for beam focusing in accelerators such as synchrotrons. In the case of electrons, the strong synchrotron electromagnetic radiation and the high synchrotron frequency (≈ 50 Hz) prohibit the use of superconducting systems. We will discuss briefly some optical characteristics of magnets which might be used for high-energy electrons.

A 1:1 magnification is produced by a long coil with a homogeneous field in the z direction. The electrons perform a screwlike motion. If the aperture angle is small ($\cos\gamma \approx 1$), the first image occurs at a distance

$$l_0 = 2\pi p/eB_z$$

where p is the momentum of the particle. The radius ρ of the trajectory is given by

$$\rho = (p/eB_z)\,\gamma$$

If the energy of the electrons amounts to 1 GeV, the distance l_0 is ≈ 5 m, with $B_z \approx 4$T. Such lenses exhibit large aberrations with

$$c_s = \pi\rho$$

$$c_c = \pi\rho$$

In order to minimize the spherical aberration. large-aperture field gradients have to be introduced.

The focal length depends in general on $\int H_z dz$ in long lenses, so the focusing power remains rather weak in lenses of this type. A series of short lenses, each with a focal length f_0 proportional to $\int H_z^2 dz$, does not bring the desired strong focusing effect, as can be easily estimated from the relation

$$f \approx 2d/\pi k^2$$

which is valid for weak lenses with bell-shaped field distributions.

After the pioneer work of Courant *et al.* (1953), quadrupole lenses have been used frequently for strong focusing in high-energy physics.

For a rough estimate we can assume the lenses to be weak, so that the relations (2.18) and (2.19) are valid. But since $f_x = -f_y$, a combination of at least two equal quadrupoles of opposite polarity placed in the proper position is necessary for convergence in every meridional plane.

power of this combination becomes, according to relations,

$$\frac{1}{f} = \frac{1}{f_{x1}} + \frac{1}{f_{x2}} + \frac{l}{f_{x1}f_{x2}} = \frac{1}{f_{y2}} + \frac{1}{f_{y1}} + \frac{l}{f_{y1}f_{y2}} = \frac{l}{f_x^{\,2}}$$

$$f_{x1} = -f_{x2} = -f_{y1} = f_{y2}$$

where l is the drift space between lens 1 and 2. The lens aberrations will not be discussed here. To demonstrate the superiority of the quadrupole lenses in regard to focusing power, let us consider an electron beam accelerated with a voltage of 10^{11} V. For a magnet with a flux density $B_p = 4$ T, a length $L_0 \approx L_1 = 0.3$ m, and an aperture $a = 0.1$ m, the focal lengths $f_x \approx -f_y$ are ≈ 1 m [equations (2.18), (2.19)]. If two quadrupoles separated by 0.5 m are used, the doublet focal length becomes 2 m. As the electron energy obtained so far with accelerators is far below 100 GeV, the conventional quadrupole lenses must not necessarily be replaced by superconducting ones in order to take advantage of the high fields and field gradients. However, there may be other reasons, such as vacuum requirements, small space, the proximity of other helium-cooled components, economic considerations especially in regard to operating costs, etc., for preferring the superconducting version.

The order of magnitude of the optical effect of dipole magnets on high-energy particles can be roughly estimated from equation (2.17). The bending angle of a 10-GeV beam passing through a transverse field 1 m long and with a flux density of 4 T will be of the order of 10°.

A.2. Magnet Designs

The superconducting magnets so far built for use with high-energy particles are iron-shrouded. Achieving mechanical stability and small tolerances is one of the main problems because of the very strong magnetic forces resulting from the high fields, because of the large dimensions, and because of the stress and strain due to the different thermal expansion coefficients of the material used.

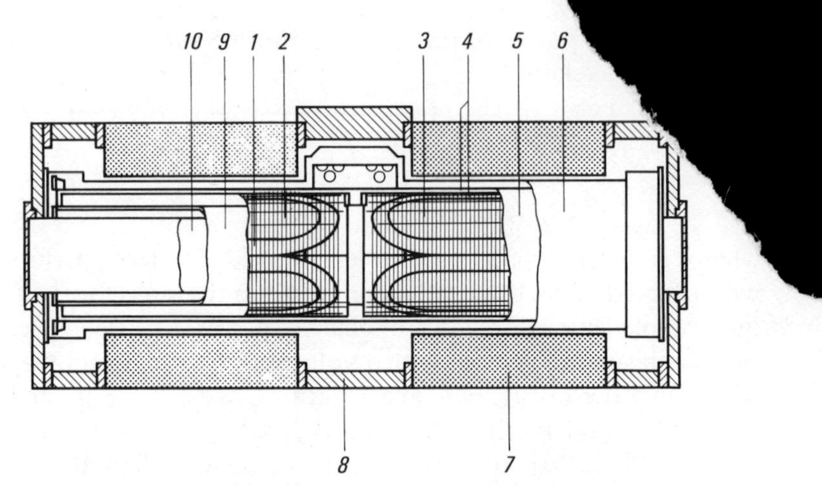

Figure A.1. Superconducting quadrupole magnet for the beam transport line of the University of California bevatron (a synchrotron producing high-energy heavy ions). Reproduced by permission of Gilbert *et al.* (1973). (1) Winding; (2) winding spacer; (3) Dacron string; (4) vacuum superinsulation; (5) liquid helium vessel; (6) liquid nitrogen cooled shield; (7) iron; (8) vacuum vessel wall; (9) liquid helium vessel inner wall; (10) room temperature tube.

The basic principle of quadrupole designs without iron pole pieces is shown in Figure 5.8. Arrangements such as type 1 of Figure 5.8 are often designed with saddle-shaped coils to decrease fringing field effects. A beam transport quadrupole doublet (Figure A.1) for the bevatron* was built and tested in the Lawrence Berkeley Laboratory (Gilbert *et al.*, 1973). Some of the data for this doublet are: bore diameter 0.25 m (warm bore 0.2 m), effective magnet length 1.03 m, peak flux density 4.5 T (gradient 28 T/m).

A few examples of designs for the superconducting iron-pole-piece free-dipole magnets are given in Figure A.2. In Figure A.2a the current is equally distributed in two intersecting circles or ellipses. The flux density within the aperture for the case of circles is given by

$$B = \mu_0 i_{\text{eff}} (R - r)$$

* The bevatron is a large synchrotron for producing high-energy protons and heavy particles.

where R is the radius of the current-carrying circle and r is the radius of the aperture. The same effect can be obtained with an inhomogenous current distribution (Figure A.2b).

Another design makes use of two elliptical saddle-shaped cylinder coils. This is the type used for the dipoles of the bevatron transport line in Berkeley (Figure A.3).

A possibility restricted exclusively to superconductors is the permanent high-fidelity storage of multipole fields in hollow cylinders (Garwin *et al.*, 1973). If a prototype magnet with the desired field configuration and a wide bore were available, any number of replicas could be made. The mechanical tolerances of the magnets with trapped flux are more or less irrelevant. To obtain suitable cylinders about 0.1 m in diameter and of arbitrary length, one may use a technique similar to that described in Section 3.2 for producing Nb–Sn sinter material (G. Lefrance

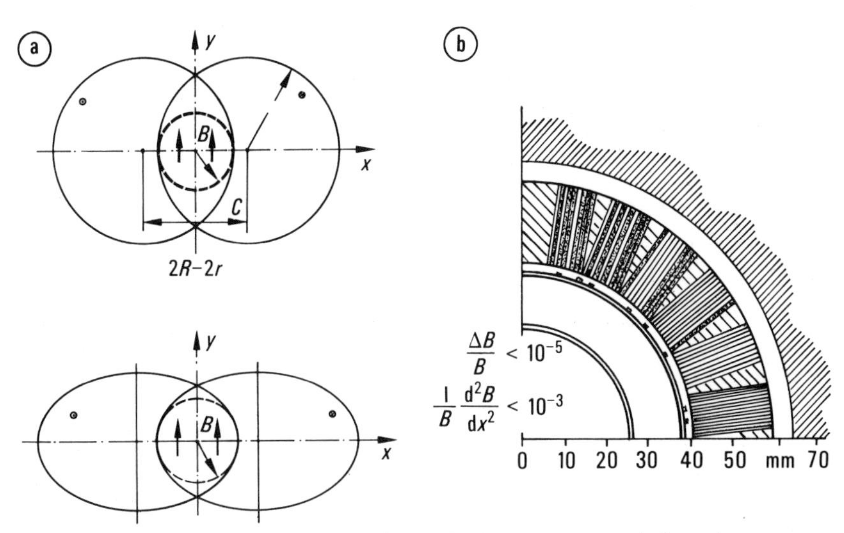

Figure A.2. Principles of large-aperture dipole magnets consisting of superconducting coils. (a) Current intersecting circles and ellipses, respectively. (b) Unequally distributed conductors. Reproduced by permission of Dahl *et al.* (1973). The current distribution corresponds to $i_{\text{eff}} \propto \cos \varphi$. The field tolerances given are calculated for an ideal magnet without magnetization effects: (▨) iron core; (▧) post and wedges; (⊟) conductors; (▤) spacers.

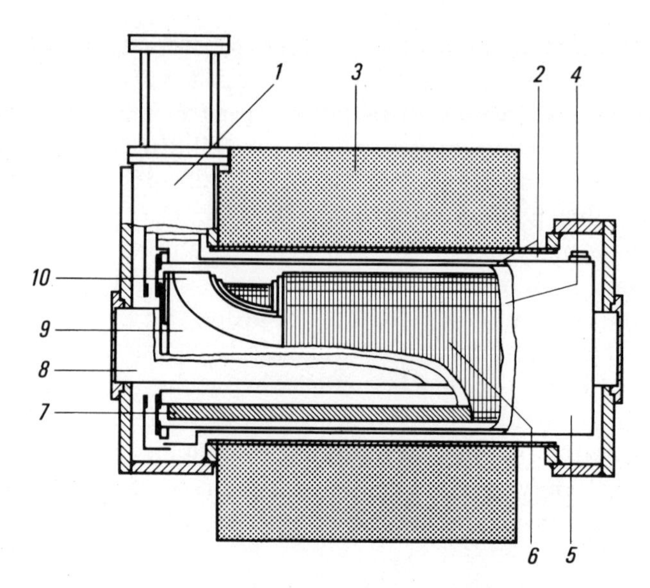

Figure A.3. Superconducting dipole magnet for the University of California bevatron beam transport line. Reproduced by permission of Gilbert *et al.* (1973). (1) Vacuum vessel; (2) vacuum superinsulation; (3) iron; (4) helium vessel; (5) liquid nitrogen cooled shield; (6) Dacron string; (7) helium vessel; (8) room temperature bore tube; (9) winding spacer; (10) winding.

and A. Müller, 1976). A rubber tube should be filled with powder with the proper ratio of the components and the tube then compressed hydrostatically. The pressed cylinder can be annealed in a pusher-type furnace after mechanical treatment. Nb–Sn sinter material of the quality described in Section 3.1 and sufficiently supported should trap fields up to at least $\mu_0 H \approx 3$ T without breaking.

B. Application of Electron Microscopy to Basic Research on Superconductivity

Electron microscopes are important tools for basic research in superconductivity, i.e., not only do we find that superconducting devices can be used to advantage in electron optical instruments, but also that electron optical instruments can be used to extend our knowledge of superconductivity.

With electron optical devices the magnetic structure in superconductors, and in some cases the motion of the structure, can be investigated. In principle, type 1 superconductors repel the magnetic field (Meissner–Ochsenfeld effect) up to the critical value H_c at which the transition to the normal state takes place. This can be derived from the magnetization curve (curve 1 in Figure B.1), yet it is only true if the demagnetization factor of the specimen is zero. Otherwise, above a certain field $H_{c1} < H_c$ the magnetization decreases (curve 2 in Figure B.1), and superconducting and normal regions, the size of which depends mainly on the geometry of the sample, are formed. The intermediate-state structure is relatively coarse as long as we consider bulk material. For this reason it can be observed with light-microscope methods.

However, type 2 superconductors (Nb, V, Tc, and most of

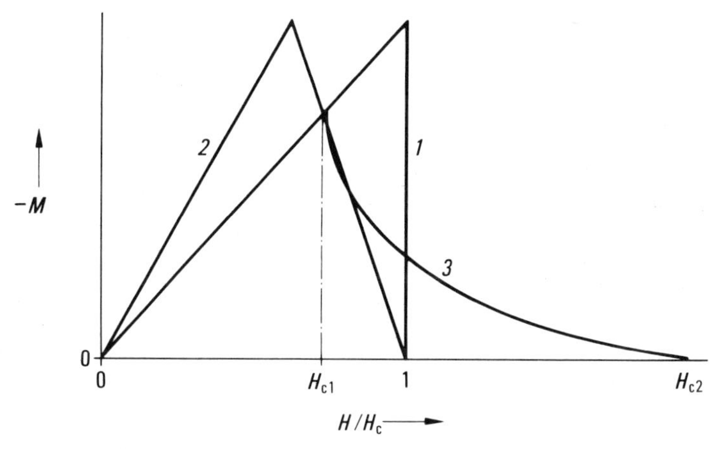

Figure B.1. Magnetization curves of superconductors: (1) magnetization curve of an infinitely long type 1 superconducting cylinder in an axial field; (2) magnetization curve of a type 1 superconducting sphere (demagnetization factor $n = 1/3$); (3) magnetization curve of a type 2 superconductor. H_e is the external field, H_c is the thermodynamic critical field which corresponds to the critical field of type 1, H_{c1} the lower critical field (transition to the mixed state), H_{c2} the upper critical field (transition to the normal state), and M the magnetization.

the alloys), some of which are of technical importance because of their high critical field (Section 3.2), exhibit total shielding up to H_{c1}. At higher fields $H_e > H_{c1}$, they are penetrated by single magnetic flux quanta,

$$\Phi_0 = \frac{e}{2h}$$

(h is Planck's constant), i.e., transition into the mixed state takes place. Thin type 1 films exhibit the same behavior (curve 3, Figure B.1).

It is assumed that in the mixed state the flux lines or vortices are arranged in a lattice with a lattice constant of 10–100 nm. In the presence of a transport current the vortex lattice is shifted as the result of Lorentz forces in an ideal type 2 superconductor. Defects in real crystalline superconductors are responsible for pinning centers, which prevent the motion of the vortex lattice to a certain degree.

In order to prove these hypotheses directly, several groups have tried to image the magnetic structure in the mixed state and thus to investigate the influence of several parameters on the kinetics of the vortices.

B.1. Imaging by the Decoration Method

The first pictures of the periodic vortex arrangement were obtained with a refined Bitter technique (Träuble and Essmann, 1966; Sarma and Moon, 1967). Small ferromagnetic particles are deposited on the superconductor in the mixed state. The particles are fixed at the points where the field gradient is highest, i.e., at the flux lines. The sample so "decorated" is warmed up and its replica investigated in the electron microscope (Figure B.2). A number of new effects have been discovered by this method, e.g., dependence of the lattice structure on the crystallographic direction of the sample that is aligned with the magnetic field, coexistence of Shubnikov and Abrikosov phases (intermediate-state and mixed-state structure; Obst 1971) (Figure B.3), etc.

This technique has recently been improved (Lischke and

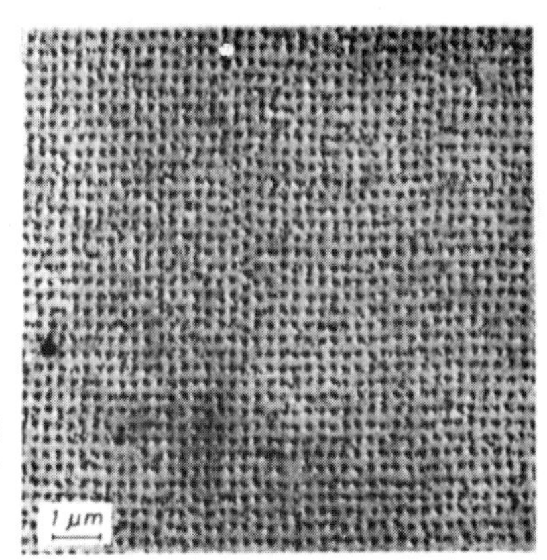

Figure B.2. Vortex lattice of a Pb/1.6 wt % Tl single crystal. Reproduced by permission of Essmann (1970). $\mu_0 H_e = 36$ mT; $T = 1.2$ K.

Rodewald, 1974; Herring, 1974), and it is now possible to examine decorated films directly in the transmission microscope. Both the vortex structure and the microstructure of the film are visible on the screen. One can thus investigate the interaction of flux lines

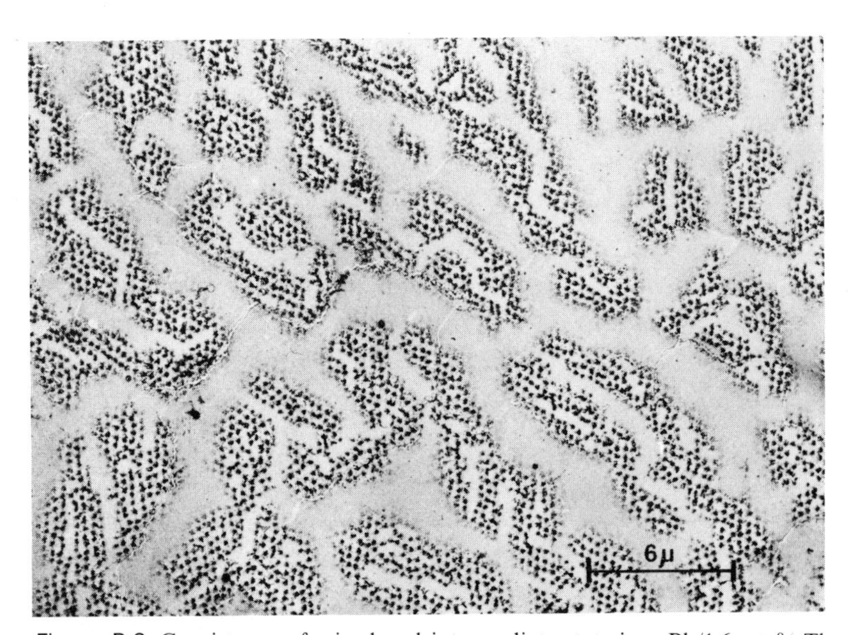

Figure B.3. Coexistence of mixed and intermediate state in a Pb/1.6 wt % Tl single crystal. Reproduced by permission of Obst (1971). Demagnetization factor $n = 0.6$; $\mu_0 H_e = 17$ mT; $T = 1.2$ K.

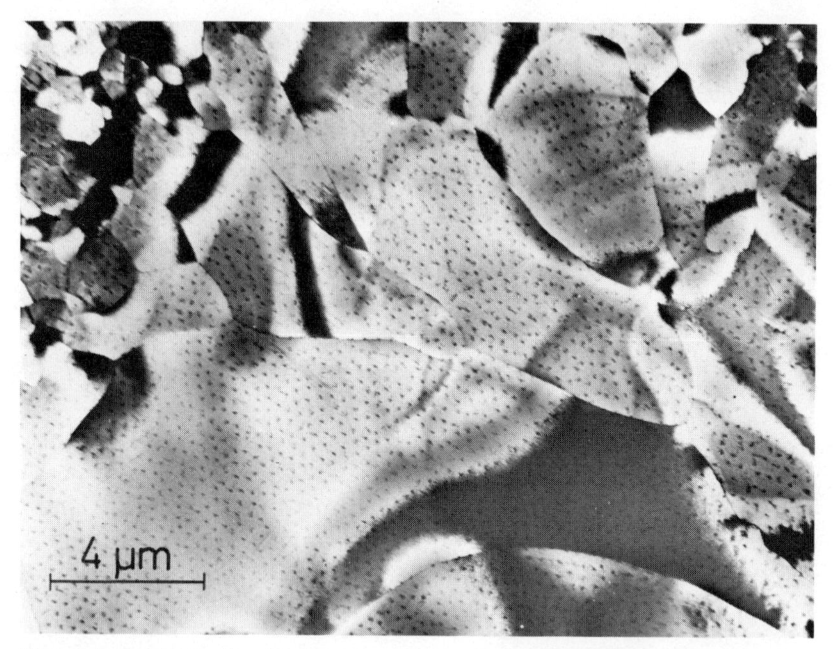

Figure B.4. Vortex distribution in a 150-nm-thick Pb film. Reproduced by permission of Lischke and Rodewald (1974). The position of the vortices is marked using the decoration method. The warmed-up specimen can be studied directly (without making a replica) in the transmission electron microscope. $\mu_0 H_e = 20$ mT; $T = 1.2$ K.

with pinning centers such as grain boundaries and precipitates (Figure B.4). However, the motion of the vortices cannot be observed.

B.2. Imaging by Electron Shadow Microscopy

Before the decoration method was developed, experiments were carried out with the superconductor cooled in a microscope liquid helium stage. Electron shadow microscopy was used to observe the field distribution along the edge of a superconductor. Either a fixed-beam method or scanning with an electron microprobe along the edge was used (Goringe and Valdré, 1964; Boersch

Figure B.5. Shadow edge of a vanadium sample above and below T_c. Reproduced by permission of Kitamura *et al.* (1966*b*). (a) T = 7 K. (b) T = 1 K. (c) T = 4.0 K. (d) T = 4.2 K.

et al., 1966*b*). The electron beam was deflected as the result of Lorentz forces due to the flux emerging from normal-conducting domains. Micrographs of the edge of a vanadium sample (critical temperature 5.1 K) exposed to small fields above and below the critical temperature are shown in Figure B.5 (Kitamura *et al.*, 1966*b*). Drastic changes of the shadow edges could be visualized. Resolution for this method on the order of magnitude of 1 μm is reported. However, an interpretation of the patterns on the micrographs is difficult, and single vortices cannot be distinguished.

B.3. Imaging by an Electron Mirror Microscope

The electron mirror method seemed a few years ago to be more promising for superconducting domain studies (Bostanjoglo and Siegel, 1967) (Figure B.6). The superconductor is the mirror electrode and is situated within the pole pieces of the lens. It is connected to the cryostat by means of a quartz single crystal for insulating the high voltage and for thermal conduction. A warm supplementary coil permits the application of fields up to $\mu_0 H_0 =$ 0.6 T perpendicular to the z azis.

The electron beam emitted from the cathode (voltage $\approx$ 25 kV) hits the mirror system after passing a bore hole in the screen. The mirror electrode is biased negatively to the cathode by a few volts, so that the electrons are reflected at a potential plane near the mirror electrode and then reach the viewing screen. Unfortunately, the resolution limit of the magnetic patterns was restricted to 10 to 20 μm, probably partly due to microroughness of the surface. For this reason only the intermediate-state structures could be studied.

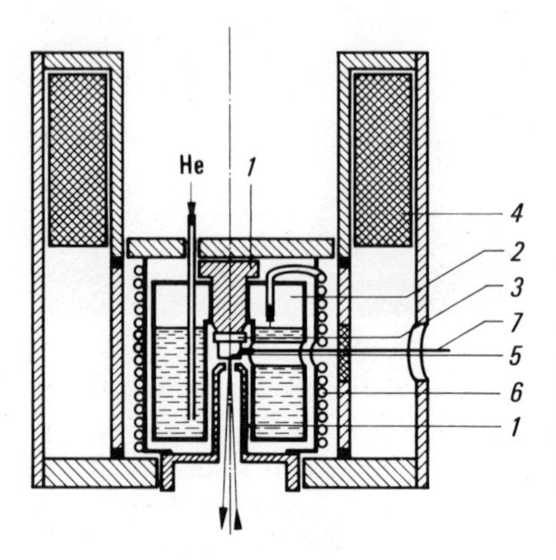

Figure B.6. Details of the mirror electrode in the mirror electron microscope. Reproduced by permission of Bostanjoglo and Siegel (1967). (1) Pole piece; (2) helium chamber; (3) quartz crystal; (4) field coil; (5) mirror electrode; (6) radiation shield; (7) high voltage.

B.4. Imaging by Lorentz Microscopy

All the efforts to investigate the mixed state in transmission with Lorentz microscopy, analogous to the imaging of ferromagnetics, where the contrast in the domain boundaries is caused by the deflection of the electrons due to the stray fields, were failures.

However, calculations of out-of-focus contrast produced by single flux quanta lying in a plane perpendicular to the beam

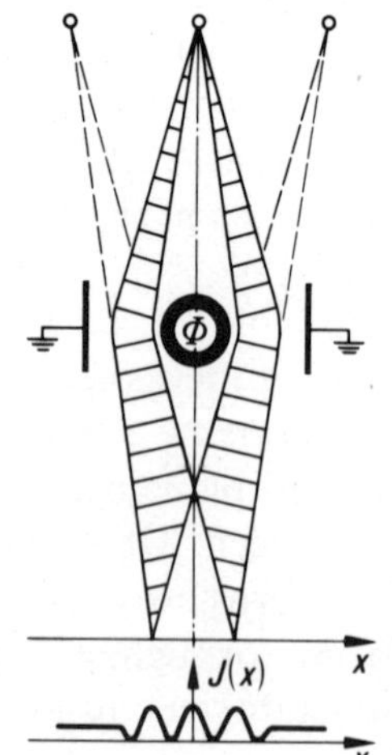

Figure B.7. Most essential part of the vortex electron microscope. Reproduced by permission of Boersch *et al.* (1974). In the center is the charged hollow superconducting cylinder with trapped flux Φ. An interference pattern is indicated below the diagram of the microscope (J is the intensity).

(Guigay and Bourret, 1967; Wohlleben, 1967) indicate that defocus fringes should be detectable. The experimental requirements for imaging a two-dimensional vortex arrangement in thin films were studied by Capilupi *et al.* (1972); they found that the conditions for obtaining sufficient contrast are rather tough, yet imaging should be just feasible.

B.5. Imaging by a Vortex Electron Microscope

Experiments are carried out for the one-dimensional case by means of a "vortex electron microscope" (Boersch *et al.*, 1974).

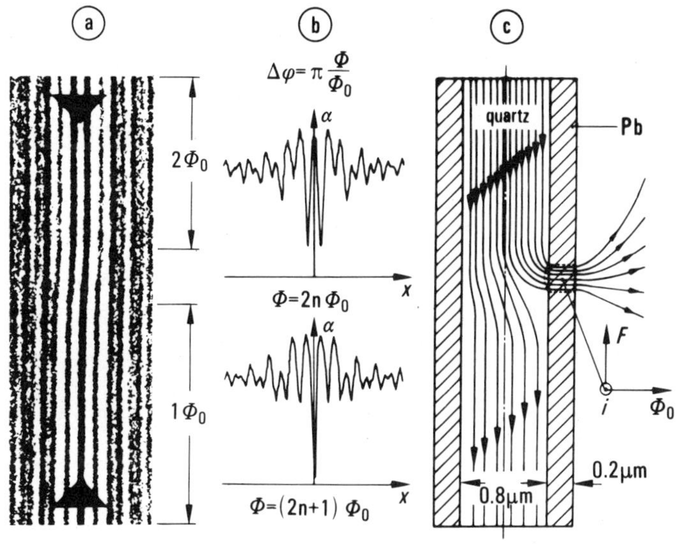

Figure B.8. Interference fringes produced in the vortex microscope by a superconducting hollow cylinder with trapped flux. Reproduced by permission of Boersch *et al.* (1974). (a) Interference pattern. (b) Densitometer curves. (c) Flux distribution in the lead cylinder as derived from (a) and (b). The flux Φ is correlated with a measurable phase shift $\Delta\varphi = \pi\Phi/\Phi_0$ (Φ_0 is the flux quantum) which can be observed in the interference fringes. The transition area from the dark ($\Phi = 1 \cdot \Phi_0$) to the bright ($\Phi = 2 \cdot \Phi_0$) symmetry center of the interference pattern determines the position of the vortex in the wall. The vortex causes a force F due to the shielding current with the density i and the flux Φ_0.

The main part of the instrument consists of an electron interferometer with an electrostatic biprism, and the Aharonov–Bohm effect (Aharonov and Bohm, 1959) is used for measurements, as shown in Figure B.7. A positively charged filament splits the electron beam. The two beams interfere below the filament and form a fringe pattern. The pattern is shifted if flux is trapped in the filament, i.e., the vector potential of the fields within the filament influences the phase of the electron waves. In practice, the biprism is installed in a liquid helium stage. The filament has the form of a hollow superconducting cylinder in which flux can be trapped (the dimensions are shown in Figure B.8). The flux distribution is reflected in the interference pattern. A change of the pattern corresponding to the region where part of the flux penetrates the wall is clearly visible in Figure B.8. Recording with a film camera permits the tracing of the temporal variations of the flux lines due to flux jumping. The model of Anderson (1962) for the temperature dependence of flux creep could be confirmed with the vortex microscope.

Some of the investigations described above would be less elaborate if the liquid helium stage and the normal objective lens were replaced by a superconducting lens.

* References

AHARONOV, Y., AND BOHM, D., 1959, *Phys. Rev.* **115**, 485.
ALBRECHT, W. W. H., 1968, *J. Appl. Phys.* **39**, 3166.
ANDERSON, P. W., 1962, *Phys. Rev. Lett.* **9**, 309.
ARCHARD, G. D., 1953, *J. Sci. Instr.* **30**, 352.
ASCHERMANN, G., FRIEDRICH, E., JUSTI, E., AND KRAMER, J., 1941, *Phys. Z.* **42**, 349.

BERGERE, R., VEYSSIERE, A., AND DANJAT, P., 1962, *Rev. Sci. Instr.* **33**, 1441.
BERJOT, G., BONHOMME, P., PAYEN, F., BEORCHIA, A., MOUCHET, J., AND LABERRIGUE, A., 1970, *in: Microscopie Électronique 1970,* abstracts of papers presented at the Seventh International Congress, held in Grenoble (P. Favard, ed.), Vol. 1, Société Française de Microscopie Électronique, Paris.
BOERSCH, H., BOSTANJOGLO, O., AND NIEDRIG, H., 1964, *Z. Phys.* **180**, 407.
BOERSCH, H., BOSTANJOGLO, O., AND LISCHKE, B., 1966*a, Optik* **24**, 460.
BOERSCH, H., BOSTANJOGLO, O., LISCHKE, B., NIEDRIG, H., AND SCHMIDT, L., 1966*b, in: Electron Microscopy 1966,* Sixth International Congress for Electron Microscopy, held in Kyoto, Japan (R. Uyeda, ed.), Vol. 1, Maruzen Co., Ltd., Tokyo.
BOERSCH, H., LISCHKE, B., AND SÖLLIG, H., 1974, *Phys. Status Solidi* B **61**, 215.
BONHOMME, P., BEORCHIA, A., AND LABERRIGUE, A., 1973, *Optik* **39**, 39.
BONJOUR, P., 1973, Thesis, L'Université de Paris Sud, Centre D'Orsay.
BONJOUR, P., AND SEPTIER, A., 1968, *in: Electron Microscopy 1968,* pre-Congress abstracts of papers presented at the Fourth European Regional Conference, held in Rome (D. S. Bocciarelli, ed.), Vol. 1, Tipografia Poliglotta Vaticana, Rome.
BOSTANJOGLO, O., AND SIEGEL, G., 1967, *Cryogenics* **7**, 157.
BOUSSARD, D., 1967, *in: Focusing of Charged Particles* (A. Septier, ed.), Vol. 2, p. 327, Academic Press, New York and London.

BRECHNA, A., 1973, *Superconducting Magnetic Systems,* Springer-Verlag, Berlin–Heidelberg–New York.

BUCHOLD, T. A., Application of superconductivity in guiding charged particles, U.S. Patent 3,008,044, Nov. 7, 1961.

CAPILUPI, C., POZZI, G., AND VALDRÉ, U., 1972, *in: Electron Microscopy 1972,* Proceedings of the Fifth European Congress on Electron Microscopy, held at the University of Manchester, The Institute of Physics, London–Bristol.

CASLAW, J. C., 1971, *Cryogenics* **11**, 57.

CODY, O., AND CULLEN, G. W., 1964, *RCA Rev.* **25**, 466.

CONTE, R. R., 1970, *Éléments de Cryogenie,* Masson and Cie, Paris.

COURANT, E. D., LIVINGSTON, M. S., AND SNYDER, M. S., 1953, *Phys. Rev.* **88**, 1168.

CREWE, A. V., EGGENBERGER, D. N., WALL, J., AND WELTER, L. M., 1968, *Rev. Sci. Instr.* **39**, 576.

DAHL, P. F., DAMM, R., JACOBUS, D. D., LASKY, C., MCINTURFF, A. D., MORGAN, G., PARZEN, G., AND SAMPSON, W. B., 1973, *IEEE Trans. Nucl. Sci.* **NS–20**, 688.

DEIS, D. W., CAVALER, J. R., HULM, J. K., AND JONES, C. K., 1969, *J. Appl. Phys.* **40**, 2153.

DIEPERS, H., MARTENS, H., SCHMIDT, O., SCHNITZKE, K., AND UZEL, Y., 1973, *IEEE Trans. Nucl. Sci.* **NS–20**, 68.

DIEPOLD, H., AND DOSSE, J., 1941, Magnetfeldlinse, Deutsche Patentschrift No. 298 216, 1941, 1953.

DIETRICH, I., 1974, *in: Proceedings of the Fifth International Cryogenic Engineering Conference, Kyoto, 1974* (K. Mendelssohn, ed.), IPC Science and Technology Press, London.

DIETRICH, I., AND WEYL, R., 1968, *in: Proceedings of the Second International Cryogenic Engineering Conference,* held in Brighton, United Kingdom, 1968, ILIFFE Science and Technology Publications, Ltd., Guildford, Surrey, United Kingdom.

DIETRICH, I., WEYL, R., AND ZERBST, H., 1967, *Cryogenics* **7**, 178.

DIETRICH, I., PFISTERER, H., AND WEYL, R., 1969, *Z. Angew. Phys.* **28**, 35.

DIETRICH, I., WEYL, R., AND ZERBST, H., 1970, *in: Microscopie Électronique 1970,* abstracts of papers presented at the Seventh International Congress, hend in Grenoble (P. Favard, ed.), Vol. 1, Société Française de Microscopie Électronique, Paris.

DIETRICH, I., KOLLER, A., AND LEFRANC, G., 1972*a, Optik* **35**, 468.

DIETRICH, I., LEFRANC, G., AND MÜLLER, A., 1972*b, J. Less-Common Met.* **29**, 121.

DIETRICH, I., WEYL, R., AND ZERBST, H., 1972*c, in: Electron Microscopy 1972,* Proceedings of the Fifth European Congress on Electron Microscopy, held at the University of Manchester, The Institute of Physics, London–Bristol.

DIETRICH, I., LEFRANC, G., WEYL, R., AND ZERBST, H., 1973, *Optik* **38**, 449.
DIETRICH, I., FOX, F., KNAPEK, E., LEFRANC, G., WEYL, E., AND ZERBST, H., 1974*a*, *in: Electron Microscopy 1974,* abstracts of papers presented to the Eighth International Congress on Electron Microscopy, held in Canberra, Australia (J. V. Sanders and D. J. Goodchild, eds.), Vol. 1, The Australian Academy of Science, Canberra.
DIETRICH, I., FOX, F., WEYL, R., AND ZERBST, H., 1974*b*, *in: High Voltage Electron Microscopy,* (P. R. Swann *et. al.,* eds.), Academic Press, London.
DIETRICH, I., HERRMANN, K.-H., AND PASSOW, G., 1975, *Optik* **42**, 439.
DIETRICH, I., FOX, F., KNAPEK, E., LEFRANC, G., NACHTRIEB, K., WEYL, R., ZERBST, H., 1976, *Optik* **44**, 469.
DOSSE, J., 1941, *Z. Phys.* **117**, 722.
DUPOUY, G., AND PERRIER, F., 1972, *C. R. Acad. Sci.* (*Paris*) **274B**, 1170.
DURANDEAU, P., AND FERT, C., 1957, *Rev. Opt.* **36**, 205.

ESSMANN, U., 1970, *in: 6. Sommerschule für Supraleitung,* held in Pegnitz/Oberfranken, Forschungslaboratorien Erlangen der Siemens AG, Munich.

FAIRBANK, W. M., PIERCE, J. M., AND WILSON, P. B., 1963, *in: Proceedings of the Eighth International Conference on Low Temperature Physics* (*LT8*), held in London, 1962 (R. O. Davies, ed.), Butterworths, London.
FERNÁNDEZ-MÓRAN, H., 1965, *Proc. Natl. Acad. Sci. USA* **53**, 445.
FERNÁNDEZ-MÓRAN, H., 1970, *in: Microscopie Électronique 1970,* abstracts of papers presented at the Seventh International Congress, held in Grenoble (P. Favard, ed.), Vol. 1, Société Française de Microscopie Électronique, Paris.
FIRTH, M., KREMPANSKY, B., AND SCHMEISSNER, F., 1971, *in: Proceedings: The Third International Conference on Magnet Technology,* held in Hamburg, 1970, Deutsches Elektronen-Synchrotron DESY, Hamburg.

GARWIN, E. L., RABINOWITZ, M., AND FRANKEL, D. J., 1973, *Appl. Phys. Lett.* **22**, 599.
GÉNOTEL, D., SEVERIN, C., AND LABERRIGUE, A., 1967, *J. Microsc.* **6**, 933.
GÉNOTEL, D., LABERRIGUE, A., PAYEN, F., AND SEVERIN, C., 1968, *in: Electron Microscopy 1968,* pre-Congress abstracts of papers presented at the Fourth European Regional Conference, held in Rome (D. S. Bocciarelli, ed.), Vol. 1, Tipografia Poliglotta Vaticana, Rome.
GILBERT, W. S., MEUSER, R. B., VOELKER, F., KILPATRICK, R. A., EATON, W. F., TOBY, F. L., AND ACKER, R. C., 1973, *IEEE Trans. Nucl. Sci.* **NS–20**, 683.
GLASER, W., 1952, *Grundlagen der Elektronenoptik,* Springer–Verlag, Wien.
GORINGE, M. J., AND VALDRÉ, U., 1964, *in: Electron Microscopy 1964,* Proceedings of the Third European Regional Conference, held in Prague (M. Titlbach, ed.), Vol. A, Publishing House of the Czechoslovak Academy of Sciences, Prague.
GUIGAY, J. P., AND BOURRET, A., 1967, *C. R., Acad. Sci.* **264**, 1389.

HAINE, M. C., AND JERVIS, M. W., 1955, *Proc. IEEE* **102B**, 265.

HANAK, J. J., 1964, *RCA Rev.* **25**, 551.

HANSSEN, K. J., AND TREPTE, L., 1971, *Optik* **32**, 519.

HARDY, D. F., 1973, *in: Advances in Optical and Electron Microscopy* (V. E. Coslett and R. Barer, eds.),Vol. 5, p. 201, Academic Press, New York.

HART, R. K., 1966, Proceedings of the AMU–ANL Workshop on High Voltage Electron Microscopy Argonne National Laboratory Report ANL-72 75.

HAWKES, P. W., 1966, *in: Quadrupole Optics, Electron Optical Properties of Rectilinear Orthogonal Systems* (P. W. Hawkes, ed.), Springer–Verlag, Berlin.

HERITAGE, M. B., 1973, *Image Processing Computer-Aided Design in Electron Optics: Proceedings,* Fifth European Congress on Electron Microscopy, 1972 (P. W. Hawkes, ed.), p. 324, Academic Press, London.

HERRING, C. P., 1974, *Phys. Lett.* **47A**, 105.

HIBINO, M., HARDY, D. F., PLOMP, F. H., KAWAKATSU, H., AND SIEGEL, B. M., 1973, *J. Appl. Phys.* **44**, 4743.

HOPPE, W., 1974, *Naturwissenschaften* **61**, 239.

HOPPE, W., LANGER, R., FRANK, J., AND FELTYNOWSKI, A., 1969, *Naturwissenschaften* **56**, 276.

HOPPE W., LANGER, R., HIRT, A., AND FRANK, J., 1970, *in: Microscopie Électronique 1970,* abstracts of papers presented at the Seventh International Congress, held in Grenoble (P. Favard, ed.), Vol. 1, Société Française de Microscopie Électronique, Paris.

HOWE, D. G., AND WEINMANN, L. S., 1974, *in: Proceedings of the Fifth International Cryogenic Engineering Conference, Kyoto, 1974* (K. Mendelssohn, ed.), IPC Science and Technology Press, London.

INOUE, K., TACHIKAWA, K., AND IWASA, Y., 1971, *Appl. Phys. Lett.* **18**, 235.

ISHIKAWA, I., OZASA, Y., KATAGIRI, S., SATO, S., AND KITAMURA, N., 1969, *in:* Soviet–Japanese Conference on Low Temperature Physics, Novosibirsk.

KAWABE, U., DOI, T., AND KUDO, M., 1969, *J. Appl. Phys.* **40**, 10.

KIM, Y. B., HEMPSTEAD, C. F., AND STRNAD, A. R., 1963, *Phys. Rev.* **129** 528.

KITAMURA, N., SCHULHOF, M. P., AND SIEGEL, B. M., 1966*a*, *Appl. Phys. Lett.* **2**, 277.

KITAMURA, N., SRIVASTAVA, O. N., AND SILCOX, J., 1966*b* *in: Electron Microscopy 1966,* Sixth International Congress for Electron Microscopy, held in Kyoto, Japan (R. Uyeda, ed.), Vol. 1, Maruzen Co., Ltd., Tokyo.

KNAPEK, E., 1975, *Optik* **41**,506.

KNAPEK, E., 1976, *Optik* **45**, in press.

LABERRIGUE, A., 1969, *J. Microsc.* **8**, 779.

LABERRIGUE, A., AND LEVINSON, L., 1964, *C. R. Acad. Sci.* **259**, 530.

LABERRIGUE, A., AND SEVERIN, C., 1967, *J. Microsc.* **6**, 123.

LABERRIGUE, A., LEVINSON, P., AND HOMO, J. C., 1971, *Rev. Phys. Appl.* **6**, 453.
LABERRIGUE, A., BERJOT, G., BONHOMME, P., GÉNOTEL, D., GIRARD, M., AND HOMO, J. C., 1974*a, in: Electron Microscopy 1974,* abstracts of papers presented to the Eighth International Congress on Electron Microscopy, held in Canberra, Australia (J. V. Sanders and D. J. Goodchild, eds.), Vol. 1, The Australian Academy of Science, Canberra.
LABERRIGUE, A., GÉNOTEL, D., GIRARD, M., SEVERIN, C., BALOSIER, G., AND HOMO, J. C., 1974*b, in: High Voltage Electron Microscopy* Academic Press, London.
LEFRANC, G., AND MULLER, A., 1976, to be published.
LEISEGANG, S., 1953, *Optik* **10**, 5.
LEVINSON, P., LABERRIGUE, A., AND TESTARD, O., 1965, *Cryogenics* **5**, 344.
LIEBMANN, G., 1949, *Proc. Phys. Soc. Lond.* **62**, 753.
LIEBMANN, G., AND GRAD, E. M., 1951, *Proc. Phys. Soc. Lond.B* **64**, 946.
LISCHKE, B., AND RODEWALD, W., 1974, *Phys. Status Solidi B* **62**, 97.
LIVINGOOD, J. J., 1969, *The Optics of Dipole Magnets,* Academic Press, New York.

MATSUDA, J., AND URA, K., 1974, *Electron.Commun. Jpn.* **57**, 67.
MERLI, P. G., 1970, *in: Microscopie Électronique 1970,* abstracts of papers presented at the Seventh International Congress, held in Grenoble (P. Favard, ed.), Vol. 1, Société Française de Microscopie Électronique, Paris.
MONTGOMERY, D. B., 1969, *Solenoid Magnet Design,* Wiley-Interscience, New York.
MÜLLER, A., AND VOIGT, H., 1970, *in: Verhandlungen der Deutschen Physikalischen Gesellschaft,* Spring Meeting, B. G. Teubner, Stuttgart.
MULVEY, T., AND NEWMANN, C. D., 1972, *in: Electron Microscopy 1972,* Proceedings of the Fifth European Congress on Electron Microscopy, held at the University of Manchester, The Institute of Physics, London–Bristol.
MUNROE, K., 1973, *Image Processing and Computer-Aided Design in Electron Optics,* p. 284, Academic Press, London.

OBST, B., 1971, *Phys. Status Solidi B* **45**, 467, 645.
OZASA, Y., KATAGARI, S., KIMURA, H., AND TADANO, B., 1966, *in: Electron Microscopy 1966,* Sixth International Congress for Electron Microscopy, held in Kyoto, Japan (R. Uyeda, ed.), Vol. 1, Maruzen Co., Ltd., Tokyo.

PASSOW, C., 1976, *Optik* **44**, 427.
PAYEN, F., AND LABERRIGUE, A., 1971, *Ann. Univ. Arers* **9**, 249.

RIECKE, W. D., 1973, *Optik* **36**, 66; 288; 375.
RIEMERSMA, H., HULM, J. K., AND CHANDRASEKHAR, B. J., 1963, *in: Proceedings of the Eighth International Conference on Low Temperature Physics* (*LT8*), held in London, 1962 (R. O. Davies, ed.), Butterworths, London.
ROSE, H., 1971, *Optik* **33**, 1; **34**, 285.
RYMER, T. B., 1966, *in:* Proceedings of the AMU–ANL Workshop on High

Voltage Electron Microscopy, Argonne National Laboratory Report ANL-72 75, p. 132.

SARMA, N. V., AND MOON, J. R., 1967, *Phys. Lett.* **24A**, 580.

SCHERZER, O., 1947, *Optik* **2**, 114.

SCHERZER, O., 1949, *J. Appl. Phys.* **20**, 20.

SCHWETTMANN, H. A., WILSON, P. B., AND CHURILOV, G. Y., 1966, *in:* Proceedings of the Sixth International Conference on High Energy Accelerators, p. 690, Rome.

SEPTIER, A., ed., 1967, *Focusing of Charged Particles,* Academic Press, New York.

SEVERIN, C., GÉNOTEL, D., GIRARD, M., AND LABERRIGUE, A., 1971, *Rev. Phys. Appl.* **6**, 459.

SIEGEL, B. M., KITAMURA, N., KROPFLI, R. A., AND SCHULHOFF, M. P., 1966, *in: Electron Microscopy 1966,* Sixth International Congress for Electron Microscopy, held in Kyoto, Japan (R. Uyeda, ed.), Vol. 1, Maruzen Co., Ltd., Tokyo.

SIEGEL, B. M., MUSINSKI, D. L., AND KUO, H. P., 1974, *in: Electron Microscopy 1974,* abstracts of papers presented to the Eighth International Congress on Electron Microscopy, held in Canberra, Australia (J. V. Sanders and D. J. Goodchild, eds.), Vol. 1, The Australian Academy of Science, Canberra.

STECKLY, Z. J. J., AND ZAR, J. L., 1965, *Trans. IEEE* **12**, 367.

STURROCK, P. A., 1951, *Phil. Trans. Royal Soc., A* **243**, 387.

SUELZLE, L. R., 1973, *IEEE Trans. Nucl. Sci.* **NS–20**, 44.

SUZUKI, S., 1965, Magnetic Objective Lens for an Electron Microscope, U.S. Patent 3173 005.

TACHIKAWA, K., AND IWASA, Y., 1970, *Appl. Phys. Lett.* **16**, 230.

TARUTANI, Y., KUDO, M., AND TAGUCHI, Y., 1974, *in: Proceedings of the Fifth International Cryogenic Engineering Conference, Kyoto, 1974* (K. Mendelssohn, ed.), IPC Science and Technology Press, London.

THON, F., 1966, *Z. Naturforsch. A* **21**, 476.

TRÄUBLE, H., AND ESSMANN, U., 1966, *Phys. Status Solidi* **17**, 813.

TRINQUIER, J., AND BALLADORE, J. L., 1968, *in: Electron Microscopy 1968,* pre-Congress abstracts of papers presented at the Fourth European Regional Conference, held in Rome (D. S. Bocciarelli, ed.), Vol. 1, Tipografia Poliglotta Vaticana, Rome.

TRINQUIER, J., BALLADORE, J. L., AND MURILLO, R., 1969, *C. R. Acad. Sci.* **288**, 1707.

TRINQUIER, J., BALLADORE, J. L., AND MATINEZ, J. P., 1972, *in: Electron Microscopy 1972,* Proceedings of the Fifth European Congress on Electron Microscopy, held at the University of Manchester, The Institute of Physics, London–Bristol.

VENEKLASEN, L. H., 1972, *Optik* **36**, 410.

VERSTER, J. L., AND KIEWIET, T. W., 1972, *in: Electron Microscopy 1972,* Proceedings of the Fifth European Congress on Electron Microscopy, held at the University of Manchester, The Institute of Physics, London–Bristol.

WEYL, R., DIETRICH, I., AND ZERBST, H., 1972, *Optik* **35**, 280.

WILHELM, M., 1972, private communication.

WOHLLEBEN, D., 1967, *J. Appl. Phys.* **38**, 334.

WORSHAM, R. E., HARRIS, W. W., MANN, J. E., RICHARDSON, E. G., AND ZIEGLER, N. F., 1974, *in:Electron Microscopy 1974,* abstracts of papers presented to the Eighth International Congress on Electron Microscopy, held in Canberra, Australia (J. V. Sanders and D. J. Goodchild, eds.), Vol. 1, The Australian Academy of Science, Canberra.

ZIEGLER, G., BLOS, B., DIEPERS, H., AND WOHLLEBEN, K., 1971, *Z. Angew. Phys.* **31**, 184.

VON ZUYLEN, P., AND FONTIJA, L. A., 1974, *High Voltage Electron Microscopy* CP. R. Swann *et. al.,* eds.), Academic Press, London.

* Index

A15 phases, 33
Aberration
  chromatic, 9, 11, 23, 24, 113, 114
    coefficients, 11-14, 49, 52, 56-58, 60, 61, 63, 89, 111
  spherical, 9-11, 14, 24
    coefficients, 11-14, 49, 52, 56-57, 60, 61, 63, 73, 89, 111
Aberration disk, 14, 21
Accelerator, 19, 95
  linear, *see* Linear accelerator, Superconducting LINAC
Anomalous transmission, 25
Astigmatism, 14, 70-72

Beam (position) detector, 98, 99, 103, 105
Bell-shaped magnetic field distribution, 7, 52, 63, 77
Bending angle, *see* Deflection angle
Bitter technique, 122
Boersch effect, 11
Brightness, 20, 91, 93, 94, 98, 102, 103
Buncher, 92, 93, 95, 98-101, 113

Carbon foil, 75, 83, 87, 88
Cardinal elements, 7, 8
  of superconducting lenses, 52, 56, 58, 63, 74
Caustic figure, 69, 72
Cavity (microwave resonator), 39, 40, 92, 93, 94, 98, 106, 114; *see also* Superconducting cavity
Coma, 14, 15, 69
Condenser lens, 6, 20, 70, 72, 77-80, 84, 86, 106-110, 112
Contamination, 25, 27, 86
Contrast, 5, 17-20, 75, 111, 112
  aperture, 6, 18
  out-of-focus, 126
  phase, 17-20
  scattering, 5, 17-20
Correction systems, 23, 61, 66-68, 72, 80-86, 89, 95, 108, 111
Critical current density, 1, 2, 33-36, 47, 48, 50-53
Critical field, 32, 33, 121, 122
Critical temperature, 31, 32, 125
Cryopumping, 25, 27
Cryostat, 26-31, 78-80, 89, 106-108
  evaporation rate of, 28, 31
  evaporator type, 27, 57
Cryotechnical material, 42, 43

Decoration method, 122-124
Deflection angle, 15, 70, 71, 117
Deflector, 20, 66, 78, 80-86, 95, 108, 109
Defocusing effect, 94, 95, 112
Dehydration, 25
Diffractogram, light optical, 75, 83, 88
Dipole field, 15, 95
Dipoles, 69, 95, 115, 118-120
Disk of least confusion, 10, 12, 21
Distortion, 15

Diverging lens, 113, 114
Drift
  in cryogenic devices, 27
  of specimen, 74, 83, 86
Duty cycle, 94

Electrolytic tank, 44, 66, 67
Electron microscope
  fixed-beam transmission mode, 5, 6, 19, 21, 97-114
  high-voltage, 25, 26, 91, 97-114
  lens testing in, 74-88
  mirror type, 125, 126
  scanning mode, 19-21, 109-114
  shadow, 124, 125
  with superconducting lenses, 77-90, 97-114
Electron optical bank, 69-70
Electron source, microwave field emission, 98-103, 112
Energy distribution of electrons, 112, 113
Energy gain of electrons, 102
Energy spread of electrons, 11, 75, 91-97, 100-102, 112-114

Ferromagnetic material, 40
  Co–Fe alloy, 41, 43, 53
  Cryoperm, 41
  dysprosium, 41, 43, 60, 61
  holmium, 41, 43, 61
  mu-metal, 25, 41
Field distribution in superconducting lenses, 44-68
Field penetration into superconductors, 38, 61, 62, 64, 70
Finite-element method, 46
Flux
  jumps, 23, 34, 36, 65, 66, 128
  lines, 122, 124, 128
  quantum, 122, 126, 127
Focal lengths
  asymptotic, 9
  of objective lens, 8-10
  of superconducting lenses, 52, 56-57, 60-63, 74, 77, 86, 107-110, 113-117
Focus, 10
Fowler–Nordheim equation, 99
Fringe system, 87, 127, 128
Fringing field, 95, 96, 105, 118

Gaussian field, 51
Gaussian function, 62
Green's function method, 46

Half-width of field distribution, 7-11, 24, 47-63
Helium, superfluid (below λ point), 27, 28, 97
High-voltage simulation device, 72, 73

Image
  aberrations, measurements of, 69-73
  reconstruction, three dimensional, 88, 89
  recording (observation), 20, 97, 111
Interferometer, electron, 128
Intermediate lens, 77-81, 86, 88, 106-111
Intermediate state (of superconductors), 121-125
Iris structure of cavity, 92, 93
Iron-shrouded superconducting lenses, 53-55, 65

Kim–Anderson relation, 34, 51, 61

Lambert distribution, 12
Laplace equation, 45
Lens strength parameter, 7-12, 52, 53, 56, 57, 63, 69, 71, 72, 74, 76, 77, 111
Lens testing, 69-76
Lenses, superconducting, *see* Superconducting lenses
Liebmann method, 13, 46, 61
Linear accelerator (LINAC), 38, 91-94, 97-104, 112-114
  duty cycle, 93, 98
  optical properties, 94
  superconducting, *see* Superconducting LINAC
Lorentz microscopy, 126, 127

Magnetic field, *see* Field
Magnetization curve, 121
Magnets
  beam-focusing, 3, 115
  beam-transport, 3, 115, 118
  energy-selection, 3

Magnets *(cont'd)*
with saddle-shaped coils, 118
of synchrotron, 115, 118
Microduty cycle, 93, 98
Misalignment, 14, 66-69, 82
Mixed state (of superconductors) 122, 123, 126
Multipole fields, 119

Octopole lenses, 24
One-field-condenser (telecentric) mode of objective lens, 9, 69-71, 76, 86, 110, 112

Persistent current
mode, 23, 36, 37, 65, 74
switches, 30, 37, 79
Phase focusing, 92, 93
Phase space, 100-102
Phase spread, 100, 101, 114
Pinning centers, 38, 122, 124
Pole pieces
iron-alloy, 56, 58, 59, 96, 104, 106
rare earth, 60, 61, 65
Principle planes, 8, 9
Projector lens, 77-79, 86, 88, 106-111

Quadrupole field, 15-17
Quadrupole focal length, 17, 117
Quadrupole magnet, 24, 85, 115-118

Radiation damage, 25
Ray equation, 13
paraxial, 13, 14
Ray tracing diagram, 8
Resolution, 11-15, 19, 21, 74, 75, 83, 87-89, 107, 110, 114
theoretical limit, 11, 12, 74, 87, 111
Resolving power, *see* Resolution

Shielding
superconducting, 2, 24, 25, 31, 57, 61, 62, 70, 80-82, 107, 122
thermal, 28-31
Shielding capability of superconductors, 34-36, 53
Shielding currents, 33, 34, 70
Shim coil, 1, 37, 53, 55

Spatial frequencies, 18, 75, 83, 87
Specimen stage
side-entry, 66, 79, 84, 108, 111
tilting, 89
top-entry, 65, 66, 79, 80, 82
Spectrometer, 95-98, 103-106, 115
energy dispersion, 105
window (picture) frame, 95, 96, 104, 115
SQUID, 106
Stigmator, 66, 80-86, 107-109
Stray fields, 19, 24, 31
Superconducting cavity, 2, 31, 38-40
Superconducting domains, 125
Superconducting elements of periodic system, 32
Superconducting lens systems, 77-89
Superconducting lenses, 23-31, 33, 45-89
coil, 23, 37, 38, 46-49, 65
iron circuit, 2, 37, 40, 53-61, 65, 72-74, 77-81, 89, 106-110
ring, 1, 2, 49-52, 65, 74, 89
shielding, 61-72, 80-89, 105-108, 110, 112
Superconducting LINAC, 2, 3, 25, 26, 94, 97-104, 112
chromatic error, 103
Superconducting material
degradation of, 3, 33, 36
Nb–Ti, 3, 34, 37, 53, 56, 77, 80, 82
Nb–Zr, 2, 34, 48, 53
Nb_3Sn, 34, 39, 48, 51, 54, 57, 59
sintered Nb–Sn, 38, 39, 53, 56, 61, 107, 119, 120
Superconducting resonator, *see* Superconducting cavity
Superconductors
high-temperature, 1, 2, 33, 34
stability of, 36
surface impedance, 39, 40
types of, 33, 121, 122
Suzuki mode of objective lens, 10
Synchronous particle, 101, 102
Synchrotron, 115, 116, 118

Transfer function, envelope of, 111, 112
Trapped flux, 33, 49, 119, 126-128

Tube diagram, 34, 35

Vibrations, mechanical, 19, 26, 27, 97, 114
Voltage stability, 74, 113
Vortex lattice, 122-124
Vortex microscope, 127, 128
Vortices, 122-128
  arrangement of, 127
  distribution of, 124

Warm iron circuit superconducting lenses, 31, 53, 54, 57, 59, 77-79